Guidebook for Radiography

Paresh R. Vaidya

# Guidebook for Radiography

 Springer

Paresh R. Vaidya
Formerly, Bhabha Atomic Research Centre
Mumbai, India

ISBN 978-981-99-8037-6        ISBN 978-981-99-8038-3   (eBook)
https://doi.org/10.1007/978-981-99-8038-3

This Springer imprint is published by the registered company Springer Nature Singapore Pte Ltd.
The registered company address is: 152 Beach Road, #21-01/04 Gateway East, Singapore 189721, Singapore

Paper in this product is recyclable.

# Preface

When the idea of preparing a guide book for radiography came, the first question we asked ourselves was, 'why one more book'? Various agencies here and abroad have published excellent books on radiography theory for the general reader as well as for the certification exams. They have their own place in the programmes of learning. However, there is a void in a specific genre—that of a book which is a ready reference for the persons appearing for the examinations. Level 3 certification presumes a certain level of proficiency with the examinees and hence there is no stipulation of a formal training course as is the case for lower levels. Only a refresher course is available. These candidates can refer to the guide book for a quick study. Certain tips not usually seen in text books are also included here. Except for a couple of chapters (e.g. Chaps. 10 and 11), it can as well be used by the candidates of Level 2 examinations also. The other practitioners of radiography can use this as a ready reference handbook for the basics.

Though most of the certification examinations have a format of multiple choice questions and do not have mathematical sums, a large number of numerical sums are included here with solutions. This is to provide the basic understanding of the underlying principles, useful in quickly solving the multiple choice questions of numerical nature.

It was the initiative of the Indian Society for Non-destructive Testing (ISNT) to get such a Guidebook published. I had pleasant interactions with Dr. P. P. Nanekar, Dr. B. Venkataraman, Dr. M. T. Shyamsundar, Mr. V. Manoharan and Dr. Bikash Ghose at different times in their capacity as the functionaries of the ISNT and benefitted from their precious feedback on the manuscript.

This is the first version of the new experiment and the readers are encouraged to communicate their feedback about its usability and suggestions for making it better.

Mumbai, India                                                       Paresh R. Vaidya, Ph.D.
August 2023

# Contents

# About the Author

**Dr. Paresh R. Vaidya** has worked with the Bhabha Atomic Research Centre (BARC) in the Department of Atomic Energy of India. He has a bachelor's degree in Physics (Honors) from Gujarat University and master's and Ph.D. degrees from the University of Bombay. His main area of work was the quality assurance of nuclear reactor fuels and has also supervised NDT of the systems in the nuclear reactor projects. He has specialized in radiography and radiometry. Dr. Vaidya has a good body of research work in the advanced areas of X-ray radiography, X-ray detection, and measurement of focal spot sizes and has over 75 national/international publications. He has contributed to the drafting and evaluation of Indian Standards on Radiography for the Bureau of Indian Standard. The author has received 'Ron Halmshaw Award' of the British Institute of NDT (BINDT), 'NDT Man of the Year' from Indian Society for NDT (ISNT) and the 'Achievement Award (R&D)' of ISNT Mumbai Chapter, R. G. Deshpande Award from 'National Association for Application of Radiation and radioisotope in Industry' (NAARRI) for propagating the use of radiation, and an Excellence Award from the Departmental of Atomic Energy, India. He is the peer-reviewer for journals of national and international repute.

# List of Figures

# List of Tables

# Chapter 1
# Basic Physics of Radiation

## Contents

All the material around us is composed of different elements and the basic building block for them all is the **Atom**. It consists of a positively charged nucleus and negatively charged electrons moving around in discreet orbits. The nucleus has two types of nucleons, viz. the chargeless **Neutrons** and positively charged **Protons**. Number of protons is equal to the number of orbiting electrons, so as to make the atom neutral, with zero net charge.

The number of Protons is denoted by '**Z**', called the 'atomic number' which is unique to each element; however the number of neutrons can vary, giving different **isotopes** of the same element. Total number of neutrons plus protons is called Mass Number **A.** Notation showing elements is like this: [$_z$E$^A$]. For example $_4$B$^9$, $_6$C$^{12}$, $_{92}$U$^{235}$ etc (Fig. 1.1).

Number of Protons in the isotope $_{12}$Mg$^{24}$ is 12. Number of Neutrons is 12.

Number of Protons in the isotope $_{12}$Mg$^{25}$ is 12. Number of Neutrons is 13.

© Ind. Society for Non-Destructive Testing 2024
P. R. Vaidya, *Guidebook for Radiography*,
https://doi.org/10.1007/978-981-99-8038-3_1

**Fig. 1.1** Picture of the
Boron atom. $Z = 4, A = 9$.
No of neutrons is 5,
electrons 4

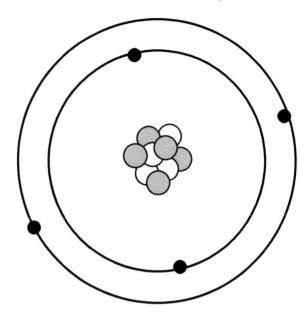

## 1.1  Radioactivity

Atoms which emit radiations are radioactive and the process is **radioactivity** or
**disintegration** or **radioactive decay**. Radioactivity can be natural or induced by
bombarding the atoms by neutrons or protons. There are three types of natural radiations $\alpha, \beta$ **and** $\gamma$. The alpha particle is same as the helium nucleus, having two positive
charges and 4 as atomic mass. $\beta$ is a negatively charged particle, with negligible mass
and $\gamma$ a neutral electromagnetic radiation, with no mass.

The unit of radioactivity in SI system is **Becquerel (Bq)**. 1 Bq is one disintegration
in a second. The older unit of **Curie (Ci)** is equal to $3.7 \times 10^{10}$ Bq. The larger units
of radioactivity are called Mega Becquerel (MBq $10^6$), Giga Becquerel (GBq $10^9$)
and Tera Becquerel (TBq $10^{12}$).

The rate of decay is called **Disintegration Constant** or **Decay Constant** $\lambda$. If $N_0$
is the number of atoms in a sample at time $T = 0$, then the atoms available for decay
at time 't' is given as

$$N = N_0 \exp^{-\lambda t} \tag{1.1}$$

This can also be written for the radiation intensity

$$I = I_0 \exp^{-\lambda t} \tag{1.2}$$

The unit of $\lambda$ can be $s^{-1}$, $h^{-1}$, day$^{-1}$ or year$^{-1}$. The time duration in which half
the atoms in a sample disintegrate is called **Half Life ($T_{1/2}$)**. $T_{1/2}$ can be in seconds,

hours, days or years, corresponding to the unit of $\lambda$. Taking $N = N_0/2$, we have $t = T_{1/2}$ and we get

$$T_{1/2} = \frac{0.693}{\lambda} \tag{1.3}$$

## 1.2 X and Gamma Rays

Both these radiations are a part of the family of Electromagnetic Spectrum. Difference is in their origin; X-rays are of atomic origin whereas gamma rays are emitted during transitions of nuclear energy levels. As per De Broglie's hypothesis, they can be considered waves as well as particles (Corpuscles). As particles they are considered the packets of energy, called **Photons**. Energy E of the photons is given as $h\nu$ or $hc/\lambda$, where $\nu$ is the frequency, h is Planck's constant, c is the velocity of light and $\lambda$ is wavelength. (This $\lambda$ is different from the decay constant mentioned above). Placing values of constants h and c, you get inverse relationship between $\lambda$ and E (Fig. 1.2)

$$\lambda(\text{in Angstrom Units}) = 12.4/\text{ Energy in keV}$$

The energy of photons and other nuclear particles is measured in **electron volt** (eV) units. It is the amount of kinetic energy gained by an electron when it travels through an electric field of 1 V. Multiples of eV are **keV** ($10^3$ eV) or **MeV** ($10^6$ eV).

$$1 \text{ electron volt} = 6.2 \times 10^{-12} \text{ ergs}$$

$$1 \text{ MeV} = 6.2 \times 10^{-6} \text{ ergs}$$

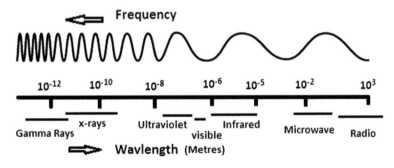

**Fig. 1.2** Electromagnetic spectrum

## 1.2.1  X-Rays

The X-ray photons are obtained by two processes viz.

(i)  Bremsstrahlung
(ii)  Characteristic X-rays

(i)  The German word ***Bremsstrahlung*** means 'braking radiation'. When a charged
     particle is abruptly accelerated or decelerated it emits radiation in the form of
     a photon. In a X-ray machine a stream of electrons is travelling between two
     electrodes and gains energy as it goes towards positively charged target/anode.
     The electron beam suddenly comes to a halt on striking the target. This produces
     photons as well as some heat. As the proportion of heat and radiation varies for
     each electron, the photons have a broad energy range. The number distribution
     of photons with various energies gives the curve shown in Fig. 1.3. The spectrum
     is termed as a **White** or **Heterogeneous Spectrum.** Maximum energy ($E_{max}$)
     here corresponds to the potential applied to the X-ray tube. Wavelength related
     to E max is $\lambda_{min}$. For example, the curve here is at 87 kV potential, so the E max
     is 87 keV.

So the related wavelength

$$\lambda_{min} = \frac{12.4}{\text{Tube Voltage in KV}} = 0.1425 \text{ A}^\circ \tag{1.4}$$

This is called Duane Hunt Limit for the wavelength.

(ii)  Characteristic X-rays

The curve in Fig 1.3 shows some small peaks superimposed on the continuous
spectrum. These are due to characteristic X-rays emitted due to bombardment of the
target material by electrons. When an inner bound electron from K, L or M—shell
is knocked out, an electron from the next shell comes to take its place. During this
process the excess binding energy is emitted as a X-ray photon; its energy is typical to
the element concerned and hence it is called the 'characteristic' X-ray. Their intensity

**Fig. 1.3**  A typical
bremsstrahlung spectrum

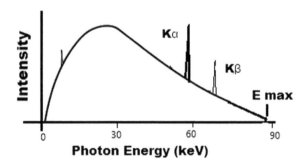

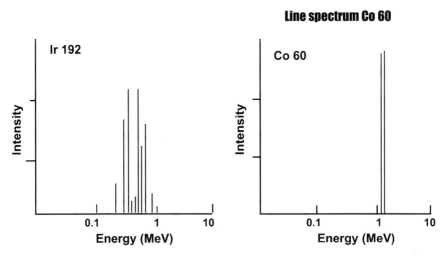

**Fig. 1.4** Line gamma spectra for Iridium-192 and Cobalt-60

is too small compared to the bremsstrahlung part to contribute to the radiography process.

### 1.2.2 Gamma Rays

These are also photons but their origin is the nucleus of the atom. In most cases a gamma photon accompanies an alpha or a beta particle emission. Unlike X-rays, the gamma rays form a line spectrum with discreet lines for each energy level. A given nuclei can have one or more energy lines. Figure 1.4 shows the line spectrum for $Ir^{192}$ and $Co^{60}$ isotopes.

## 1.3 Ionization

This is a process by which an atom or a molecule acquires a negative or positive charge by gaining or losing one electron. Photon or a charged particle when passing through material, knocks the electrons out and produces ions. Hence they are called ionizing radiations. The liberated electrons can cause further ionization. This phenomenon in gases is used for measurement of radiation quantity. Detectors like Geiger Mueller Counter etc. will be discussed in Chap. 13.

## 1.4  Radiation Measurement Units

Intensity of radiation is given as the number of particles incident on a unit area in unit time. However, for practical purposes the intensity of x and gamma rays is measured by their capability to produce ionization in the gases and is called **Exposure**. It is defined as the quantity of x or gamma radiation which produces ions carrying 1 C of electric charge of either sign in one kg of air at STP.

Thus SI Unit of Exposure is **C/kg**.

In older CGS units it is called **Roentgen** and defined as that quantity of radiation which produces 1 esu of charge of either sign in 1 cc of air at STP.

$$1R = 2.58 \times 10^{-4}\, c/kg$$

This is valid till about 3 MeV of energy.

**Dose**: Exposure delivers the energy to the medium but only what is absorbed, does the harm. Thus the dose or the **absorbed dose** is defined as the energy absorbed per unit mass of the matter. The unit is J/kg or **Gray (Gy)**, where 1 J/kg is 1 Gy. The older unit is rad, which is equal to absorption of 100 ergs of energy in one gram of material.

$$1 \text{ Gy} = 100 \text{ rad}$$

**Equivalent Dose**: Different types of radiations have different effect on biological tissues. For example, taking gamma photon as 1, alpha particles are 20 times more harmful and neutrons 5 times. These factors are called **Weighting Factors** or Quality factors or Relative biological effectiveness (RBE).

$$\text{Dose Equivalent} = \text{dose from a radiation} \times \text{its Weighting factor}$$

When more than one type of radiation is involved, Dose Equivalent will be a weighted sumof all.

$$\text{Dose Equivalent} = \sum D_R W_R$$

Unit for this is also J/kg but the special name given is **Sievert (Sv)**
Older unit was called **rem** and was quantitatively equal to rad. **1 Sv = 100 rem**

## 1.5  Inverse Square Law

Intensity of radiation from a point source reduces as the inverse of square of distance from the source. i.e.

$$\frac{I_1}{I_2} = \frac{d_2^2}{d_1^2} \tag{1.5}$$

where I is intensity and d is the distance

## 1.6  Numericals

**Example 1**  Disintegration constant $\lambda$ for Thulium-170 is 0.00537 day$^{-1}$. Find its half life.
  *Solution*: Half life $= 0.693/\lambda = 0.693/0.00537 = 129$ days

**Example 2**  Find disintegration constant for Co-60, for which half life is 5.3 years.
  *Solution*: $\lambda = 0.693/$ half life $= 0.693/5.3 = 0.13$ year$^{-1}$.
  If $\lambda$ is required in (day$^{-1}$) units, divide the value in (year$^{-1}$) by 365

$$0.13 \text{ year}^{-1} = 0.13/365 = 3.610^{-4} \text{ day}^{-1}$$

**Example 3**  What fraction of original activity will be for a source after 4 half lives?
  *Solution*: After each half life activity becomes ½ of original. Hence after 4 half lives it will be.
  $(1/2)^4$ times i.e. $1/16$ times.
  In % terms this will be $100/16 = 6.3\%$ of original.

**Example 4**  If a 10 Ci (370 GBq) source of Ir 192 delivers an exposure of 5 R at 1 m distance, what will be the exposure at 3 m distance?
  *Solution*: In the formula of Eq. 1.5, I1 is the intensity at one metre distance. $I_2$ is the intensity at 3 m distance.

$$\text{Hence } 5R/I_2 = 3^2/1^2$$

$$\text{i.e.  } I_2 = 5/9 = 0.56\,R$$

Eventually a formula will not be required to be used. One can directly use inverse square relation and divide 5 by $3^2$.

**Example 5**  Radiation intensity from a curie of Co-60 source is 1.3 R/h at 1 m. At what distance the exposure rate will be 2 R/h?
  *Solution*: By inverse square Law $I_1/I_2 = D_2^2/D_1^2$. $I_1$ is 1.3 R/h. $I_2$ required is 2 R/hr. Hence $1.3/2 = D_2^2/1$, and $D_2 = (1.3/2)$ ½ ~ 0.8 m Ans.

**Model Questions**

Q.1   Number of protons in a nucleus is same as

(a)   Number of neutrons in the nucleus
(b)   Atomic weight of the isotope
(c)   Number of electrons in the orbit
(d)   All the above

Q.2   Two isotopes of a given element do not have

(a)   Same atomic number
(b)   Same chemical symbol
(c)   Same mass number
(d)   Same chemical properties

Q.3   Ionization in a gas is related to the mass and charge of the radiation causing the ionization. Which of he following will generate largest number of ions?

(a)   Neutrons
(b)   Alpha Particles
(c)   Gamma rays
(d)   Electrons

Answers are available in the section "Answers to the Model Questions".

# Chapter 2
# Radiographic Equipments

## Contents

Conventional radiography is carried out using X-rays and gamma rays. Neutrons are also used for radiography. Neutron radiography technique is discussed in Chap. 9.

## 2.1   X-Ray Units

A typical X-ray machine consists of an X-ray tube, a high voltage generator and the controls. In portable units the tube and the generator could be located in single chamber called Tube-head. X-ray tube is typically a diode where filament is cathode and disc shaped target is mounted on the anode. The heated filament emits electrons by the **thermionic emission** process. Under the applied voltage they go towards the disc shaped target fixed on the anode, causing Bremsstrahlung.

(Special constructions like microfocus unit could be like a triode where the target could be different from anode. Refer Chap. 9).

Intensity of X-rays is proportional to atomic number of the target, tube current and square of the applied voltage.

© Ind. Society for Non-Destructive Testing 2024
P. R. Vaidya, *Guidebook for Radiography*,
https://doi.org/10.1007/978-981-99-8038-3_2

$$I = C i Z V^2 \qquad\qquad (2.1)$$

where C is a constant with value between 1 to $3 \times 10^{-9}$.

As more than 99% of energy is lost as heat, cooling of target and anode becomes necessary. This could be done using gas, water or oil.

Thus the target selection involves 3 factors:

- Higher atomic number (for higher conversion efficiency)
- High melting point (to avoid local burn out or damage due to heat)
- Better thermal conductivity (to avoid local heating and resultant burn out)

This makes tungsten a natural choice, among others like Molybdenum. The envelope of X-ray tube is made of glass or ceramics. It is sealed at high vacuum and a port called **Window** is provided to get the useful beam of X-rays out. This is usually made of light materials like Be or Al to permit transmission even at low kV. The window reduces the intensity and alters the spectrum nominally depending upon its own thickness. This is called the **inherent filtration**.

## 2.2  Control Circuits

There are different types of electrical circuits to produce high voltage to be applied between anode and cathode. Figure 2.1 shows various voltage wave-forms generated by customized circuits called Villard, Graetz and Greinacher circuits. Most basic form is the 'self-rectified circuit' where X-ray tube itself is working as a diode and gives 'half wave' voltage form. Others respectively produce half wave (continuous), full wave and constant potential (CP) wave forms. Modern circuits use inter-mediate frequency or high frequency power supply to obtain smoother waveforms in CP units.

The X-ray units are either **Bipolar** or **Unipolar**. If the cathode is negative and anode is at ground potential, it is still positive with respect to cathode and attracts electrons. Similarly cathode at ground and anode positive also works. Both these modes are called Unipolar. When Cathode is connected to negative generator and the Anode to positive generator, the unit is called Bi-polar.

## 2.3  X-Ray Parameters

Control unit has a provision to select three parameters of operation viz. the tube voltage (kV), tube current (mA) and a Timer to set Exposure time. If the unit is a dual focus unit, it will also give a choice to select big or the small focal spot.

*Tube Voltage (kV)* determines the energy of the beam. Thicker objects or objects made of higher Z, need higher kV.

*Tube Current (mA)* determines the flux (number of photons), usually called the beam intensity.

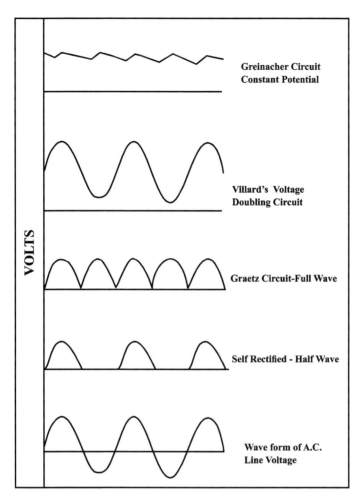

**Fig. 2.1**  Output voltage waveforms for different type of X-ray units

*Exposure* is measured as (time × milliamps). e.g. exposure time of 12 min at 5 mA tube current means the exposure of 60 mA-min.

*Focal Spot* is an important feature of a tube and decides the resolution of the radiograph. A smaller focus gives sharper images and a larger focal spot can deliver higher tube current.

Figure 2.2 shows how the spectrum changes with the change in tube voltage and tube current. Note that the peak position shifts to higher energy when kV is increased. When **only** mA is increased the peak position and E max remains same. The higher tube current demands better cooling of the target/anode; if that can not be provided the unit should be put off for some time every exposure cycle. Proportion of time the unit keeps ON during an operational cycle is called **Duty Cycle**. 50% duty cycle means it can be operated for half the time in a given period. If it could be operated

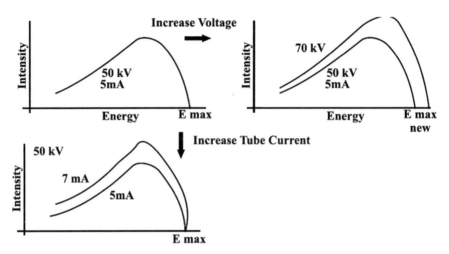

**Fig. 2.2** How spectrum changes with changes in kV, mA

**Table 2.1** kV, mA and the nature of X-ray beam

|            | Lower mA                   | Higher mA                      |
|------------|----------------------------|--------------------------------|
| Low kV     | Soft X-rays, low intensity | Soft X-rays, higher Inensity   |
| Higher kV  | Hard X-rays, low intensity | Hard X-rays, high intensity    |

continuously without a cooling break it is called 100% duty cycle. Real Time RT units need to be operated this way (Table 2.1).

### 2.3.1  Heel Effect

The maxima of the X-ray intensity (or flux) is not exactly at the point below the target but a small distance away towards the cathode on anode–cathode axis. This phenomenon is called the Heel Effect.

### 2.3.2  High Energy Equipments

For tube voltage higher than 450 kV the coil sizes become big and transformer losses are high. A special transformer called Resonant transformer was used to minimize the losses, but this has gone out of use now. Instead, accelerator based X-ray sources viz. **Van de Graff generator, Linear Accelerator** and **Betatron** are used. Refer Chap. 9 for more on these.

## 2.4 Radio-Isotope Sources

The source equipment for gamma ray radiography is called **Industrial Radiography Exposure Device** or popularly, a camera. It is made of an efficient shielding material like lead, tungsten alloy or depleted uranium. It is operated remotely by a flexible cable at the end of which the pig tail assembly is connected. The actual isotope source is in a capsule at the end of the pig tail. The conduit tube housing the pig tail in the camera is of 'S' shape to prevent streaming of radiation.

Isotopes with reasonably long half life and energies in the range required for industrial RT are used as the gamma sources. All except Radium and $Cs^{137}$ are produced by irradiation in the reactor by the process called '**activation**'. Radium is a naturally occurring substance and was used initially. Due to better alternatives it has been phased out. $Cs^{137}$ is acquired from fission products from the nuclear fuel. It is not used for industrial RT anymore as it comes in powder form as Cesium Chloride and hence is a potential health hazard (Table 2.2).

Typical nuclear reactions for the activation of all other nuclei are as follows:

$$_{27}Co^{59} + _0 n^1 \text{ (neutron)} \rightarrow _{27} Co^{60}$$

$$_{27}Co^{60} \rightarrow \beta + _{28} Ni^{60} + \gamma$$

$$_{77}Ir^{191} + _0 n^1 \rightarrow _{77} I^{192} \rightarrow \beta + _{78} Pt^{192} + \gamma$$

$$\text{Or by electron capture} \rightarrow _{76} Os^{192} + \gamma$$

$$_{34}Se^{74} + _0 n^1 \rightarrow _{34} Se^{75} \rightarrow \text{electron capture} \rightarrow _{33} As^{75} + \gamma$$

The value RHM, also called **Dose Rate Constant**, depends upon the energy of the output radiation of the source.

**Table 2.2** Properties of isotopic sources

| Isotope | Half life | Energy | RHM[a] R/h/Ci | Nominal Fe thickness tested |
|---------|-----------|--------|---------------|------------------------------|
| $_{88}Ra^{226}$ | 1590 years | 0.6, 1.12, 1.76 meV | 0.825 | – |
| $_{55}Cs^{137}$ | 30 years | 0.66 meV | 0.37 | 20–100 mm |
| $_{77}Ir^{192}$ | 74 days | 0.31, 0.47, 0.64 meV | 0.5 | 19–65 mm |
| $_{27}Co^{60}$ | 5.3 years | 1.17, 1.33 meV | 1.3 | 40–150 mm |
| $_{69}Tm^{170}$ | 127 days | 52 keV, 0.87 meV | 0.0025 | 2.5–12 mm Al:10–30 mm |
| $_{34}Se^{75}$ | 120 days | 0.066, 0.265, 0.135, 0.4 meV | 0.2 | 8–30 mm |
| $_{70}Yb^{169}$ | 31 days | 0.17, 0.2 meV | 0.125 | 3–12 mm Al: 8–40 mm |

[a]**RHM** is the term used for radiation field in Roentgen per Curie per hour at the distance of 1 m from the source

**Specific Activity**: Radioactivity per gram of isotope material. This depends upon the activation cross section ($\sigma$) of the irradiated substance, neutron flux $\Phi$, time the source was activated and the decay constant of the isotope produced. The maximum (theoretical) value it can achieve is

$$S = \lambda N_0/A (Bq/g) \tag{2.2}$$

$N_0$ is the Avogadro Number and A atomic weight. The higher specific activity makes the physical size of the source smaller, which reduces the self absorption and also gives a sharper image.

## 2.5 Decay Chart and Curve

Radiographer needs to know the activity of the source periodically in order to decide the exposure time. It is not convenient to use the formula $N = N_0\ e^{-\lambda t}$ so often. Instead a table of reduction factor or a decay curve is used. Table 2.3 gives fraction of original activity after the lapse of time shown. This fraction is to be multiplied with the activity at time zero. For example if a 8 Ci Co-60 source was purchased 2 years ago, its activity today will be $0.77 \times 8 = 6.16$ Ci.

Each isotope has its own decay curve. Figure 2.3 gives the decay curve for Ir 192 and Co-60 sources. Figure 2.3a is exponential whereas Fig. 2.3b is seen linear because it is drawn with the logarithmic y-scale. The residual activity falls to half for Ir-192 every 74 days and every 5.3 years for Cobalt. If in Fig. 2.3a, the x-axis were to be marked with "Number of Half Lives" instead of multiples of "74 days", the same curve can be used for any isotope.

**Table 2.3** Decay Fractions

| Cobalt 60 Half life 5.3 years | | Iridium 192 Half life 74 days | |
|---|---|---|---|
| Age (years) | Factor | Age (weeks) | Factor |
| 0.5 | 0.94 | 1 | 0.94 |
| 1 | 0.88 | 2 | 0.88 |
| 1.5 | 0.82 | 3 | 0.82 |
| 2 | 0.77 | 4 | 0.77 |
| 2.5 | 0.72 | 5 | 0.72 |
| 3 | 0.67 | 7 | 0.63 |
| 3.5 | 0.63 | 9 | 0.55 |
| 4 | 0.57 | 11 | 0.49 |
| 4.5 | 0.55 | 13 | 0.43 |
| 5 | 0.51 | 15 | 0.38 |

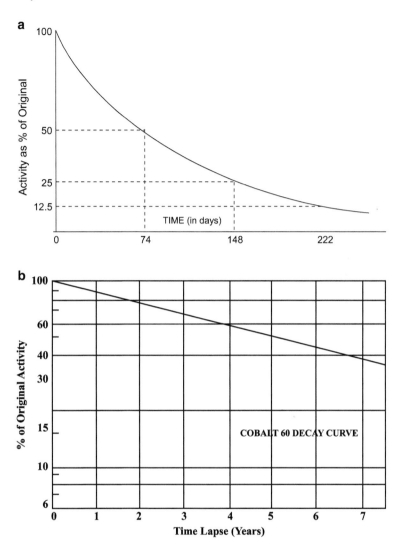

**Fig. 2.3** **a** Decay curve for Ir 192 source, **b** Decay curve for Co-60 source

## 2.5.1   *Comparison of X-Rays and Isotope Sources*

| X-rays | Gamma rays |
|---|---|
| 1. Need electricity for operation | No electricity required |
| 2. Energy (kV) can be changed | Energy is fixed for an isotope |
| 3. Difficulty in reaching inaccessible spots | Can be fixed for inaccessible spots |

(continued)

(continued)

| 4. Radiation hazard limited | Radiation protection to be ensured carefully |
|---|---|
| 5. —— | Decay Correction required to find source strength |
| 6. Beam Intensity fairly high | Usually beam intensity low |
| (Exposure times short) | (Usually Long exposure times) |

> **Nugget:**
> Spectrum for a CP unit is harder than that for a self-rectified Unit for the same kV operation. This is because the self-rectified wave form includes lower voltage component while reaching the maximum value. The same is true for decreasing part of the spectrum. Thus the average or RMS value of voltage is lower than the peak. Hence the lower effective energy for such units for same peak value.

## 2.6   Numericals

*Example 1* Working with a Co-60 source, a radiographer waiting at a distance of 3 m receives a dose of 96 mrem in 10 min. What is the source strength?

*Solution*: For 1 Ci source his dose will be $= \text{RHM} \times 1/3^2 \times 1/6 = \text{RHM}/54$

($1/3^2$ for distance correction for 3 m and 10 mins $= 1/6$ h)

RHM for Co-60 is 1.3 r/h hence dose will be 1.3/54 rem or 1300/54 mrem $= 24$ mrem

24 mrem comes from 1 Ci, but he received 96 mrem. Therefore the source must have been

$$96/24 \text{ Ci} = \textbf{4 Ci}.$$

(For information: 100 mrem $= 1$ mSv. Seivert Sv is the SI Unit for dose)

*Example 2* Adopt Fig. 2.3a for Cs-137 isotope by marking the x axis with its half life. Find activity of a 10 Ci (370 GBq) source after 100 years.

*Solution*: Not that the y-axis in Fig. 2.3a is in Percent of the Original activity. Thus it can represent Cs decay if the x-axis is modified. Half life of Cs 137 is 30 years. On the graph replace 74, 148 etc days by 30, 60, 90 and 120 years.

100 years on x-axis of this linear graph will correspond to about 10% on y- axis. Hence the 10 Ci would become 1 Ci after 100 years. [Verify same by calculation. Fraction will be inverse of $2^{(100/30)}$].

**Model Questions**

Q.1  An industrial X-ray unit output consists of

(a)  Bremstrahlung radiation
(b)  Photons of same energy
(c)  Characteristic-rays
(d)  (a) and (c)

Q.2  Three X-ray tubes have used three different target materials viz. copper, tungston and aluminium. If all other parameters and geometries are identical, the output X-ray intensity will be in decreasing order as follows

(a)  Tube with copper target, with aluminium target, with tungsten target
(b)  Tube with aluminium target, with tungsten target, with copper target
(c)  Tube with tungston target, with copper target, with aluminium target
(d)  Tube with tungston target, with aluminium target, with copper target

Q.3  Dose rate constant (or RHM) of an isotope source depends upon

(a)  Strength of the source
(b)  Energy of the source
(c)  Diameter of the source pellet
(d)  None of these

Q.4  The purpose to use Beryllium (Be) material as the window on X-ray tubes is

(a)  To obtain high output from x- ray tube
(b)  To retain lower kV photons in the beam
(c)  Be is not suitable as a window material due to low Z
(d)  Both (a) and (b)

Q.5  Two X-ray machines operating at same nominal kV and mA setting, will give

(a)  The same intensities and energies of radiation
(b)  The same intensities but produce different energies of radiation
(c)  The same energies but may produce different intensities of radiation
(d)  Not only different intensities, but also different energies of radiation

Answers are available in the section "Answers to the Model Questions".

# Chapter 3
# Interaction of Radiation with Matter

## Contents

The predominant interaction of x and gamma photons with the material is at the atomic level and very rarely at the nuclear level. They are energy packets which affect the electrons in the orbits. Three processes which can occur are:

- Photo electric effect
- Compton Scatter
- Pair Production.

## 3.1 Photo Electric Effect

An electron in the orbit **closest** to the nucleus is knocked out and all the energy of the photon is consumed in the process (Fig. 3.1). Any energy left out will become kinetic energy of the electron ejected. If an another electron from the next orbit takes place of the knocked electron, there will be an emission of a characteristic X-ray photon. Probability of the P.E. Effect is higher for the low energy photons and in high atomic number (Z) materials. It is proportional to $Z^m/E^n$. Typical value in some study is $\tau \propto \frac{Z^5}{E^{3.5}}$.

© Ind. Society for Non-Destructive Testing 2024
P. R. Vaidya, *Guidebook for Radiography*,
https://doi.org/10.1007/978-981-99-8038-3_3

**Fig. 3.1 a** Photo electric
effect **b** Compton effect
**c** Pair production

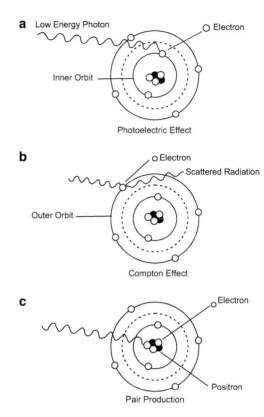

## 3.2   Scattering

Rayleigh or Elastic scattering does not change the energy of photon and even the
path change is mainly in forward direction. Its influence is limited to very low kV.
But the inelastic, Compton scattering has a bearing on radiographic process. Photon
knocks out an electron form the **outer orbit** of the atom and its own direction of
travel changes after the collision. The process occurs at intermediate energies, used
prominently in radiography (RT). Its probability (σ) increases linearly with atomic
no. Z and decrease with the energy of incoming radiation, but not as fast as for PE
effect. It has negative influence on RT because the direct beam going to the film (or
detector) gets mixed with the scattered photons which do not represent image data.
Thus a reduction in contrast.

**Fig. 3.2** When radiation
travels through specimen

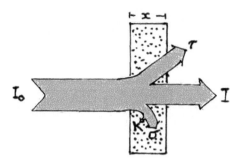

## 3.3 Pair Production

This occurs at energies > 1.02 meV, in presence of a nucleus. A pair of electron and
positron is produced whereas the photon disappears. Majority of RT operations are
not affected by this as it occurs at higher energies. Probability denoted by 'κ'.

## 3.4 Attenuation Coefficient (μ)

All the three processes seen in the previous section reduces the intensity of the
incident beam before it comes out on the other side of the specimen. The fraction
of intensity lost during the travel through a unit length of material is called **Linear
Attenuation Coefficient**. It is the sum of reductions due to PE effect, Compton scatter
and pair production Fig. 3.2.

$$\mu = \tau + \sigma + \kappa \text{ (respective probability of the three processes)} \qquad (3.1)$$

$\tau$ and $\kappa$ are true absorptions whereas $\sigma$ only changes the direction and deviates the
beam. Thus the term **Attenuation Coefficient** is better than **Absorption Coefficient**,
though both are used. Like its components, $\mu$ also depends upon Z and energy of
radiation. One approximation (by Bragg and Pierce) is

$$\mu = k\frac{Z^3}{E^3} \qquad (3.2)$$

constant k depends upon material density.

## 3.5 Build Up Factor

If $\mu$ is the fraction lost in a unit distance, the fraction lost in dx distance

$$dI/I = -\mu dx$$

Integrating and finding constants, we get

$$I = I_0 \exp(-\mu x) \tag{3.3}$$

The dimension of $\mu$ is $(\text{Length})^{-1}$. But this is true only in ideal conditions when the beam of radiation is very narrow. For the broad beam geometry the scattered radiation adds to the output intensity and the Eq. 3.3 takes the form

$$I = BI_0 \exp(-\mu x) \tag{3.4}$$

where B is called Build Up Factor.

$$B = \frac{I \, \text{direct} + I \, \text{scattered}}{I \, \text{direct}}$$

Or,

$$B = 1 + \frac{I_s}{I_d} \tag{3.5}$$

B is always $\geq 1$ and often as high as 5. Its value depends upon the specimen thickness, energy and beam profile including use of diaphragm, mask etc.

Substitute $I = \frac{1}{2} I_0$ in Eqs. 3.3 or 3.4, you get value of thickness **x** where the beam intensity falls to half its original. This thickness is called **Half Value Layer (HVL)** and works out to be

$$HVL = \frac{0.693}{\mu} \tag{3.6}$$

Material thickness at which the intensity falls to one tenth of original is termed as Tenth Value Layer (**TVL**).

$$TVL = 3.322 \, HVL \tag{3.6a}$$

HVL is also called Half Value Thickness (**HVT**) and TVL called **TVT**.

## 3.6   Effective $\mu$

$\mu$ is a function of energy and the value in Eq. 3.3 will be for a particular energy. The X-ray spectrum used in RT is a white, continuous spectrum containing all energies in the range $0-E_{max}$. Attenuation Coefficient $\mu$ for this composite beam will be a weighted average of large number of $\mu$ values representing all energies in the beam.

This is called **Effective $\mu$ ($\mu_{eff}$)**. In practice this can be empirically determined by obtaining **average effective HVL**, and then using Eq. 3.6 to find $\mu_{eff}$.

## 3.7 Mass Attenuation Coefficient ($\mu_m$)

So far we have discussed the linear attenuation coefficient. The $\mu$ in Eq. 3.6 is actually $\mu_L$. The term $\mu x$ is dimensionless. For linear x, absorption coefficient is $\mu_L$. If x is given in mass terms (i.e. aerial density units, g/cm²), then the absorption coefficient will be called **Mass Attenuation Coefficient $\mu_m$** (or **Mass Absorption Coefficient $\mu_m$**) and its unit **cm²/g.** If $\rho$ is the density of material in g/cc units then

$$\mu_m = \mu_L/\rho \quad \text{or} \quad \mu_m\rho = \mu_L \tag{3.7}$$

Figure 3.3 shows the relationship between $\mu_m$ and energy. The broad nature of graph for linear coefficient also is similar.

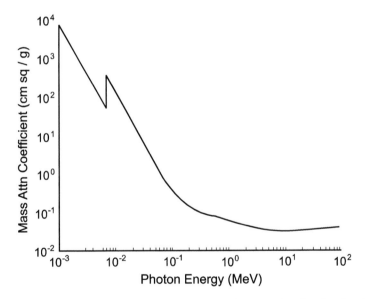

**Fig. 3.3** Typical mass attenuation coefficient as a function of energy. (Based on data from National Institute of Standards and Technology, USA on their website https://www.nist.gov/pml/x-ray-att enuation-coefficients)

### 3.7.1  $\mu_m$ of a Compound

Mass Attenuation Coefficient $\mu_m$ is useful in finding absorption in a combination of materials. There are two situations.

 (i)  If the composition of the material by weight is known, or
(ii)  If the chemical formula of the compound is known.

 (i)  For example, one can find attenuation by Cupro-Nickel material if one knows $\mu_m$ for copper and nickel individually. Let us assume the alloy is Cu–Ni (70:30). Call mass absorption coefficient of Cu as $\mu_{m(cu)}$ and that for Ni as $\mu_{m(ni)}$. Then the absorption coeff. of the composite material at that energy is

$$0.7\mu_m(cu) + 0.3\mu_m(ni).$$

If you like to get the linear $\mu$ for the practical application, get the density of composite material and then use Eq. 3.7.

(ii)  In case of a chemical formula like $A_nB_k$ (just as $CaCl_2$ or $Al_2O_3$), the composite coefficient is

$$\mu_m = \frac{\mu_{m(A)}.a.n + \mu_{m(B)}.b.k}{an + bk} \tag{3.7a}$$

where a and b are the atomic weights of the element A and B, n and k integers for valencies. Take $Al_2O_3$ as an example; its mass absorption coefficient is

$$\mu_m = \frac{\mu_{m(Al)} \times 27 \times 2 + \mu_{m(0)} \times 16 \times 3}{(27 \times 2) + (16 \times 3)} \tag{3.7b}$$

Values for $\mu_m$ for Aluminium and Oxygen have to be found from the tables for the energy of interest. $\mu_m$ can be converted to linear coefficient if the density of the material is known.

## 3.8  Shielding and Collimation

Materials with high $\mu$ are used as radiation shielding in different applications. Lead is the most commonly used material as it is amenable to easy forming in different shapes. Tungsten is costly and difficult to fabricate; hence it is used for the collimation purpose where quantity needed is not very large. Concrete does not have high $\mu$ but can be used in bulk economically, thus is used in walls of enclosures having high radiation levels. Figure 3.4 gives a chart relating concrete thickness with attenuation. Relationship between thickness and absorption is exponential but the curve is a straight line as it is plotted on a semi-log scale.

**Fig. 3.4** Attenuation for
different thicknesses of
concrete

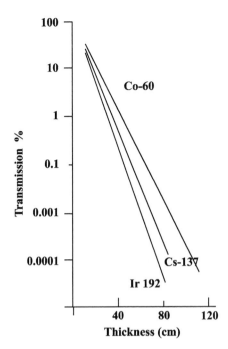

## 3.9 Numericals

*Example 1*  HVL of Lead (Pb) for Ir 192 source is 5 mm. Find its linear attenuation
coefficient.

Solution: $\mu_L = 0.693/HVL = 0.693/5 = 0.1386mm^{-1}$ i.e.$1.386cm^{-1}$.

*Example 2*  Scattered intensity is 3 times the direct intensity in an RT set up. What
is the value of build up factor?

Solution: Using Eq. (3.5), B = 1 + Is/Id but Is = 3Id

Therefore B = 1 + 3 = 4

*Example 3*  If the build up factor is 8, what is the component of scattered radiation?

Solution: Using Eq. (3.5), 1 + Is/Id = 8

Therefore Is/Id = 7 or $I_{scattered} = 7 I_{direct}$

*Example 4*  Mass Attn Coeff for Cu and Ni at 200 keV energy are respectively 0.156
and 0.1582 cm²/gm. Find $\mu_L$ for Cu–Ni 70:30 grade alloy. Density for Cu–Ni is
8.9 g/cc

Solution: Composite Attenuation Coefficient for Cu–Ni (70:30) is 0.7 weightage
for $\mu_m$ of copper and 0.3 weighttage for $\mu_m$ of nickel.

$$\text{Composite } \mu_m = 0.7\mu_{m.cu} + 0.3\mu_{m.Ni}$$
$$= 0.7(0.156) + 0.3(0.1582)$$

$$= 0.1567 \text{ cm}^2/\text{g (Mass Attn Coefficient)}$$

Now to find linear attn. coefficient for the alloy, use Eq. (3.7).

Linear $\mu_{LcuNi} = 0.1567 \text{cm}^2/\text{g} \times 8.9 \text{g/cm}^3 = \mathbf{1.394 \, cm^{-1}}$ at 200 KeV

**Example 5** At 100 keV the Linear Attn Coefficients for Cu and Ni Are 4.0935 and 3.907 cm$^{-1}$ Respectively. Find Linear Coeff of Cu Ni (90:10) Alloy at that Energy.

(Specific Gravity of Cu 8.93 g/cc, for Ni 8.8 g/cc, for Cu–Ni 8.94 g/cc)

*Solution*: As the composite $\mu$ can be found only for Mass Absoption coefficients, first get the $\mu_m$ for both the materials.

Using Eq. (3.7), $\mu_m = \mu_L$/Sp gravity.

Hence $\mu_{m.cu} = 4.0935/8.93 = 0.458 \text{ cm}^2/\text{g}$

$$\mu_{m.Ni} = 3.907/8.8 = 0.44396 \text{ cm}^2/\text{g}$$

$$\text{composite } \mu_m = 0.9 \times \mu_{m.cu} + 0.1\mu_{m.Ni} = \mathbf{0.4566 \, cm^2/g}$$

Now to find linear attn. coeff. For the alloy use Eq. 3.7 again

$$\mu_{LcuNi} = \mathbf{0.4566} \times 8.94 = 4.082 \text{ cm}^{-1} \text{ at 100 KeV}$$

**Example 6** Find Linear Abs Coeff $\mu L$ for Aluminum Oxide ($Al_2O_3$) with Density = 3.5 g/cc at Energy of 100 keV x-rays.

Given that $\mu_m$ for Al = 0.17 cm$^2$/gm and $\mu_m$ for Oxygen is = 0.155 cm$^2$/gm at 100 keV.

*Solution*: Use Eq. (3.7a) to find $\mu_m$ of alumina first.

$$\mu_m = \frac{\mu_{m(Al)} \times 27 \times 2 + \mu_{m(0)} \times 16 \times 3}{(27 \times 2) + (16 \times 3)} \qquad (3.7c)$$

Put values of $\mu_m$ for Al and Oxygen here.

$$\mu_m = \frac{0.17 \times 27 \times 2 + 0.155 \times 16 \times 3}{(54 + 48)}$$

$$= \frac{9.18 + 7.44}{102} = 0.1629 \text{ cm}^2/\text{gm}$$

$\mu_L$ for $Al_2O_3$ is 0.1629 × 3.5 (density of alumina) = 0.57 cm$^{-1}$. **Ans**

**Nugget**

In a thick object, successive HVLs are larger and larger as you go in the material.
Hence 1st HVL < 2nd HVL < 3rd HVL and so on.

This is because the lower energies are filtered and the radiation beam keeps hardening as it travels deep inside the material. Thus the next HVL is higher.

**Model Questions**

Q.1 Which of the following statements is true?

(a) Pair production will be dominating effect during RT of 5 mm steel plate.
(b) Compton effect is prominent effect encountered during RT at intermediate energies.
(c) Scattering has no effect on the contrast in radiography.
(d) Photo-electric effect takes over during Ir-192 radiography

Q.2 The Build Up Factor for radiation passing through a material is given by the expression

(a) $I_s/I_d$
(b) $\frac{I_s+I_d}{I_d}$
(c) $\frac{I_s+I_d}{I_s}$
(d) $I_d/I_s$

Where $I_s$ = Scattered Intensity and $I_d$ = Direct beam intensity.

Q.3 An inclusion with x-ray absorption coefficient $\mu_i$ lies in the material of the coefficient $\mu_m$. The appearance of the inclusion will depend upon

(a) $\mu_i + \mu_m$
(b) $|\mu_i - \mu_m|$
(c) $\mu_i \cdot \mu_m$
(d) None of these

Answers are available in the section "Answers to the Model Questions".

# Chapter 4
# Radiographic Films and Screens

## Contents

Radiographic film is composed of a cellulose-acetate or a polyester base with a layer of photo sensitive emulsion material on **both** sides of the base. Emulsion is silver halide (largely silver bromide, a small amount of silver iodide) suspended in gelatin. Figure 4.1 shows a typical structure. A thin bonding or subbing layer sticks the emulsion to the base. A super coat (or top coat) of hardened gelatin is to protect the film from mechanical damage.

When the radiation falls on the film, it brings a physical change in the emulsion which is called the **Latent Image**, because it can not be observed by the naked eye. Photon knocks out the electron from the bromine anions and frees bromine from the AgBr lattice. Process of latent image formation can be explained on the basis of band theory of solids and electron trap sites as per the **Gurney Mott theory**. The latent image can be made 'visible' if subjected to the chemical processing. Film processing has essentially four steps as development, stop bath, fixing and rinsing.

**Developer** is an alkaline solution which acts on the grains having latent image. It is a reducing agent and supplies electrons to silver ions to convert into silver atoms, Thus a fine layer of transparent metallic silver forms which is the visible image. The developer contains one or more of the following compounds viz. pyrogallol, hydroquinone, metol etc. No immediate action happens on the unexposed grains, without a latent image.

© Ind. Society for Non-Destructive Testing 2024
P. R. Vaidya, *Guidebook for Radiography*,
https://doi.org/10.1007/978-981-99-8038-3_4

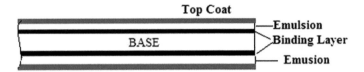

**Fig. 4.1** Structure of a X-ray film

**Stop Bath** is meant to arrest the development action immediately before proceeding to fixer solution.

**Fixer** is usually sodium thiosulphate solution in acidic environment. It washes away the unexposed and undeveloped silver halide compound as also 'fixes' silver atoms by removing bromide ions.

**Hardener** makes the supercoat tough to avoid scratches on the films. It is added to fixer in manual processing and to developer for automatic processing.

**Rinsing** takes away residual chemicals from the emulsion to increase the shelf life of the radiographs.

**Drying** is the last step in the processing sequence.

   *Storage* of radiographs (as well as raw stock films) is to be done in dry atmosphere at 20 °C temperature to avoid deterioration.

## 4.1   Film Density

The degree of blackness of film is measured as the optical density (D). It is given as

$$D = \log_{10} I_d/I_t \tag{4.1}$$

where $I_d$ is he light incident on the film and $I_t$ is light transmitted through it.

$$\text{In other words} \quad D = \log_{10}\left[\frac{1}{transmittance}\right] \tag{4.1a}$$

   For $D = 1$, 1/10th of incident light will be able to come through the film. At $D = 2$, it is 1/100.

   Developer action will be higher at the higher temperatures giving higher density. Thus there is a need to control processing temperature.

## 4.2 Types of Films

There are three factor defining quality of a film

(i)   Grain Size
(ii)  Speed
(iii) Contrast.

Grain size decides the resolution or definition. Speed and grain size are linked as the higher grain size collects the exposure more efficiently and thus gets black faster. In principle contrast is due to chemical properties of the emulsion. But incidentally for radiographic films, it is linked with the grain size. Fine grain films have higher contrast and large grain films have lower contrast. Thus there are 3 varieties of films in the market

Slow films—they are fine grain films with high resolution and high contrast
Medium speed films—have medium grain size and medium contrast and resolution
Fast films—have larger grain size and relatively lower contrast and resolution
Screen type films—This is an obsolete concept in industrial radiography. These films were designed to be used with fluorescent screens.

Films were divided into three **classes** called class 1, 2 and 3 respectively from slow to fast speeds and having high to low contrast. However this qualitative classification is now removed from ASTM standard. Instead a new concept of 'film system class' is brought in; the '**film system**' includes the recommended processing chemicals and the performance classification is based on a quantitative measure using the value of Gradient G and Granularity $\sigma_D$ (see ASTM E-1815).

**Film Factor**: It is the exposure required to obtain density 2 on a given film. Faster the film, lower is the Film Factor. This is not a universal parameter because the FF changes with the energy of radiation. That is why the concept is more popular in Isotope radiography than for X-rays.

## 4.3 Characteristic Curve

It is also called **Sensitometric Curve** or Herter and Driffield Curve (**H & D Curve**). It depicts the relationship between the exposure to the film and the optical density obtained (see Fig. 4.2). As the exposure range covered is large, the x-axis shows the logarithm of relative exposure. The H & D curve is specific to the film type, density, source to film distance (SFD) and development process. Sometimes the values of relative exposure are plotted using the graph scale which is logarithmic (semi log paper) as shown in Fig. 4.3. **Note this aspect carefully as it can be useful during Numericals on this subject.**

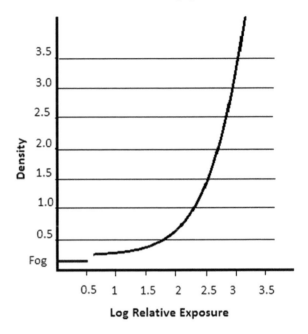

**Fig. 4.2** Film characteristic curve

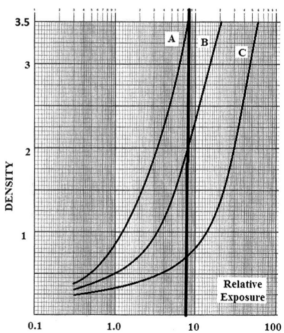

**Fig. 4.3** Comparison of films for speed and contrast

### 4.3.1   Use of Characteristic Curve

(i)   *To Compare Film Speed and Contrast*

H & D curve can be used to compare speeds of different films; for this more than one curve needs to be drawn on the *same* graphs as shown in Fig. 4.3. Here the same exposure of 8 units (where a line is drawn) yields the density of 3.5 for film 'A' whereas films B and C reach only density 1.9 and 0.7 respectively. Thus the speed of **film A > speed of B > speed of C.**

Slope of the H & D Curve is called **Gradient** and it indicates the contrast of the film. Speed and Contrast can get affected by the type of developer as well as time and temperature of processing.

$$\text{Gradient G} = \frac{d(D)}{d(\text{Log rel Expo})} = \frac{D_2 - D_1}{\log \text{rel } E_2 - \log \text{rel } E_1} \qquad (4.2)$$

Gradient can be *at a given density* by drawing a tangent at that point and taking its slope. Or it could be an *average gradient* between two densities. Both are shown in Fig. 4.4. The lower triangle shows gradient at the point where density is 0.8 and it is $\frac{a}{b}$ (about 1.4). Average Gradient is shown between densities 2 and 3, given as PQ/QR. This is about 5. It also shows that the gradient of a H & D curve increases with the density.

(ii)   *Correcting the Exposures*

**Fig. 4.4**   Two ways of taking gradient

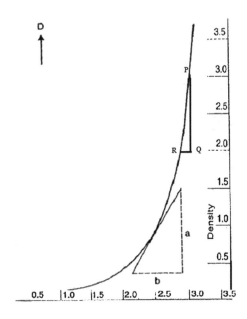

Another use of this curve is in adjusting the exposure for under or over-exposed films. Assume that you need to achieve density 3.5 on the film in Fig. 4.2, but achieved only 2.0 by giving the exposure of 30 mA min. What should be the new exposure?

Find log relative exposure for both these densities.

$$\text{Log rel exp for D } 3.5 = 3.05 \qquad (4.2a)$$

$$\text{Log rel exp for D } 2.0 = 2.75 \qquad (4.2b)$$

(a) − (b) = 0.3. Take antilog of (0.3). This is the ratio of new and old exposures.
Antilog (0.3) × 30 mA min = new exposure.
Antilog 0.3 = 2, hence new exposure is 60 mA min to achieve density 3.5.

**Practical tip**: *One way to find antilog is to use standard Log Tables.*
*Other way is to find on calculator as $10^x$ (**Antilog $x = 10^x$**). For example (a − b) in above example is (0.3). On a calculator use the button $[x^y]$ to find $10^{0.3}$. Button sequence is like this: [10] $[x^y]$ [0.3]. Screen will show the answer 1.995, which is nearly 2.*

## 4.4  Intensifying Screens

Primary function of these screens is to reduce the exposure time. Photoluminuos materials like ZnS and $CaWO_4$ were used earlier, called Salt Screens. Their Intensification Factor (IF) i.e. reduction in exposure time was high, more than 100 times. But they bought down the resolution because of spread of light inside the emulsion. Hence now Lead (Pb) intensifying screens are used. Intensification here happens due to photoelectrons and not light, thus retaining the resolution. Their IF is only 2–3 but they filter scattered radiation and thus improve contrast. These are not to be used below the radiation energy of 100–120 kV as they absorb the radiation and exposure time increases rather than decrease.

Fluorometallic screens use combination of Pb and $CaWO_4$, taking good things from both, viz better resolution and high IF (∼ 10 times). Characteristic curve does not change using the Pb screens but changes when using Salt screens or Fluorometallic screens.

### 4.4.1  Fog Density

Even the unexposed film can have some non-zero density on it when developed. This can come from ageing, radiation exposure during storage, exposure to visible light, fumes of chemicals etc. (Fog density for film C in Fig. 4.3 is 0.25).

### 4.4.2 Film Graininess

This is a subjective condition caused by the statistical fluctuation in the photon numbers absorbed and number of developed grains in a unit area of the film. Its nature depends upon the emulsion properties (thickness, grain density, size distribution etc.) and radiation energy. Energetic photons release secondary electrons within emulsion, which in turn make some more grains developable by their interaction, which are not related to original image. This adds to blur and is called **Inherent Unsharpness** of the film $U_i$ or $U_f$, because it can not be eliminated. It increases with energy of radiation and is about 0.3 mm at 1 MeV energy.

**Difference between photographic and radiographic films**

| Photo film | X-ray film |
|---|---|
| Thin layer of emulsion ($\sim \mu$) | Thick layer ($\sim 10 \mu$) |
| Finer grains than X-ray film | Larger grains in 3 grades |
| Emulsion on one side of film | Emulsion on both sides |
| Some photons per grain needed | One photon per grain to make it developable |
| Development in full darkness | Red and Green safe lights permitted |

## 4.5 Numericals

Use the curve given in Fig. 4.2 to solve Examples 1 and 2.

***Example 1*** Radiography of a specimen covers density range 0.9–2.0. If lower density 0.9 is raised to 1.5, what will be the highest density on the film?

*Solution*: First find the ratio of exposures for 1.5 D and 0.9 D. (We will use this notation D to indicate Optical Density for brevity.) Apply that ratio to 2.0 D to find new Density.

From the above curve, Log rel exp for 0.9 D = 2.25
Log rel exp for 1.5 D = 2.5 hence $\Delta$ log rel exp = 2.5 − 2.25 = 0.25

(we do not need anti log of 0.25 because absolute value of exposure is NOT required).
Add this to log rel exp of 2.0 D, which is 2.75
$$2.75 + 0.25 = 3.0 \quad \text{On y-axis this corresponds to density 3.3.}$$
**Thus highest density which was 2.0 D will now be 3.3 D.**

***Example 2*** A particular RT job needed 1 h exposure using 3 Ci Co-60 source to get 2.25 D. However the operator was not aware that the source was actually 4 Ci. What density will he actually achieve?

*Solution*: Source being of higher activity, exposure became 4/3 (i.e. 1.33) times higher.

$$\text{Log } 1.33 = 0.1239.$$

This needs to be added to the Log rel exp value corresponding to 2.25 D which is 2.75. $2.75 + 0.1239 = 2.8739$ is the value of log rel exp of new density.

Referring to the Curve it gives **2.75** on the density axis. **That will be the density actually achieved**.

*Example 3*  Repeat Example 2 for film C in Fig. 4.3.

*Solution*: X-axis in Fig. 4.3 is 'Relative Exposure' but the scale is logarithmic instead of linear. Hence no need to take Log of ratio 1.33 obtained in Example 2, but directly multiply the rel exp value for 2.25 D with 1.33.

Rel Exp for 2.25 D is 33 on film C.

$33 \times 1.33 = 43.89$ which is the actual rel exp value. On the graph, this corresponds to density 2.85. **Actual density achieved will be 2.85.**

*Example 4*  RT of a job yielded density 3.5 on Film A (Fig. 4.3) at 30 mA min exposure. However needed density is only 2.35. What should be the new exposure?

*Solution*: Rel Exp is 8 at 3.5 D

Rel Exp is 4 at 2.35 D.

As the graph is on Semi-log paper, **not difference but ratio** is to be taken of two values $8/4 = 2$.

Hence the required exposure is $30/2 = 15$ mA min.

(If a doubt comes whether to divided by the ratio or multiply, check if the exposure needed will be higher or lower. Here, obviously the new exposure will be less as the density required is lesser than achieved. Hence **divide** by the ratio factor.)

*Example 5*  A weld was radiographed on film B (Fig. 4.3) using an exposure of 5 Ci hrs. Find the exposure if film C is to be used for the same job. Density obtained at weld is 2.0.

*Solution*: Rel Exp at density 2 for film C is 30

Rel Exp at density 2 for film B is 8.

Film C is slower and will need higher exposure by the factor $30/8 = 3.75$.

(This is the speed ratio of two films at 2.0 D).

Exposure on film C is $3.75 \times 5$ Ci hrs $= 18.75$ or ~ 19 Ci hrs.

*Example 6*  Find average gradient between densities 2.5 and 1.5 for film C in Fig. 4.3.

*Solution*: Rel Exp at 2.5 D is 39, at 1.5 D is 22.

$$\text{Gradient G} = \frac{D_2 - D_1}{\log \text{ rel } E_2 - \log \text{ rel } E_1}$$

As the graph is on Log paper, what we read out is not already the log (rel exp). Thus logarithm needs to be taken 'relative exposure' values obtained from x-axis.

$Log\ 39 = 1.591$ and $Log\ 22 = 1.3424$

$$log\ rel\ E_2 - log\ rel\ E_1 = 1.591 - 1.3424 = 0.2485$$

$D_2 - D_1 = 1, \therefore G = 1/0.2485 = $ **4.024 is the gradient** between 1.5 and 2.5 D.

(Note that the log scales are NOT linear; remember this fact while making visual estimates of distance between sub division markings.)

## Model Questions

Q.1 Lead foil screen are used in industrial radiography

    (a) To reduce the exposure time

    (b) To improve the image quality by reducing the scattered radiation on film

    (c) Both a and b are valid reasons

    (d) Neither a nor b are reasons to use Pb screens

Q.2 The processed and dried radiographic film has a thin black semi transparent layer. It is made of

    (a) Silver bromide grains

    (b) Silver atoms

    (c) Silver Halide particles

    (d) Silver ions

Q.3 What is true about the latent image?

    (a) Latent image gets developed by the fixer solution

    (b) Photons and electrons have important role to play in its formation

    (c) Latent image also forms in fluorescent screens

    (d) Can be viewed in dark room before development

Answers are available in the section "Answers to the Model Questions".

# Chapter 5
# Basics for Radiography Techniques

## Contents

## 5.1 The Technique Chart

**R**adiography produces a 2-dimensional projection (or a shadow) of the 3-dimensional object using X-rays or gamma rays. In a typical RT set-up the source, its size and the specimen thickness are fixed, whereas Source to Focus distance SFD (some times called FFD in case of X-rays), the type of film and the exposure parameters (kV, mA, time or Ci-hrs) can be controlled by the radiographer. To get the correct radiograph with proper density a **Technique Chart** is used. It is also called **Exposure Chart** (Fig. 5.1). It gives the exposure value for the thickness being tested. This chart is valid for the particular X-ray machine or the isotope source. Out of many options of kV, the one selected is such that the desired thickness falls on the mid zone of the line. Many Codes give guide lines for selection of the kV. Following parameters are also fixed for the chart

- Material being tested
- Film/screens and processing
- SFD
- Density.

© Ind. Society for Non-Destructive Testing 2024
P. R. Vaidya, *Guidebook for Radiography*,
https://doi.org/10.1007/978-981-99-8038-3_5

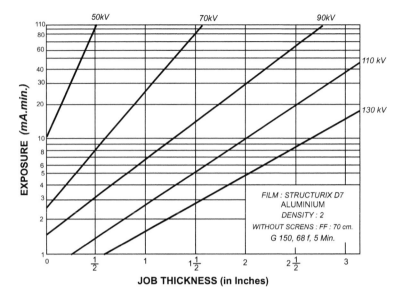

**Fig. 5.1**  Technique chart for X-rays for aluminium

However in practice if any of the factors is different from that given on the chart, one can make corrections to re-calculate the exposure as shown below.

*Material* Technique chart is usually made for steel or aluminum. Exposure times for other materials can be computed using the Equivalence Factors given as Table 5.1. Occasionally even the kV may need to be changed if materials have vastly different atomic number and density. Radiographic Equivalence Factors are essentially derived from the ratio of linear coefficient $\mu_L$ of both the materials. Exact method of using these factors is given in Numericals section of this chapter.

*Film* Film manufacturers provide figures for relative speeds of their range of films. If the film used is faster (or slower) the exposure time obtained from the chart can be divided (or multiplied) by the ratio of relative speeds.

*SFD* The chart in Fig. 5.1 is made at 70 cm. Inverse Square law is used to modify the time obtained from the chart for any other SFD. However Eq. (1.5) in Chap. 1 i.e. $I_1/I_2 = D_2{}^2/D_1{}^2$ is meant for intensity; it will change the form for the exposure time. Exposure (or exposure time) is inversely proportional to intensity of radiation. Thus it is **directly** proportional to distance.

$$\frac{E_2}{E_1} = \frac{D_2^2}{D_1^2} \quad \text{and} \quad \frac{t_2}{t_1} = \frac{D_2^2}{D_1^2} \tag{5.1}$$

**Table 5.1**  Radiographic Equivalence Factors

| | X-rays kV | | | | | Gamma rays | |
|---|---|---|---|---|---|---|---|
| | 50 | 100 | 150 | 220 | 400 | Ir 192 | Co 60 |
| Magnesium | 0.6 | 0.6 | 0.05 | 0.08 | | 0.22 | 0.22 |
| Aluminium | 1.0 | 1.0 | 0.12 | 0.18 | | 0.34 | 0.34 |
| Titanium | | 8.0 | 0.63 | 0.71 | 0.71 | 0.9 | 0.9 |
| Steel | | 12 | 1.0 | 1.0 | 1.0 | 1.0 | 1.0 |
| Copper | | 18 | 1.5 | 1.4 | 1.4 | 1.1 | 1.1 |
| Zinc | | | 1.4 | 1.3 | 1.3 | 1.1 | 1.0 |
| Brass | | | 1.4 | 1.3 | 1.3 | 1.1 | 1.1 |
| Lead | | | 14 | 11 | | 4.0 | 2.3 |
| Zirconium | | | 2 | 1.9 | 1.5 | 1.3 | 1.2 |
| Uranium | | | | 20 | 12.5 | 12.5 | 3.4 |

*Note* Upto 100 kV, reference material is aluminium

Thus the inverse square law takes a new form of **Direct Square Law** for exposure and exposure time.

*Density* If density other than 2 is required, H & D curve is used to make correction on the exposure obtained from the chart.

Sample examples for the upgrades due to these four parameters are given in the Numericals section of this chapter.

For gamma radiography also exposure charts exist (see Fig. 5.5). However some people use formula given here as a direct method to calculate the exposure.

$$\text{Exposure time (h)} = \frac{Film\ Factor \times SFD^2 \times 2^{\frac{t}{HVL}}}{Ci \times RHM} \qquad (5.2)$$

where SFD in metres, t is the specimen thickness.

## 5.2  Geometric Factors

Refer to Fig. 5.2 showing the geometry of the exposure set up. If $D_0$ is source to object distance, the distance from Source to Film (SFD) is '$D_0 + t$'. For X-rays this is also called Focus to Film distance (FFD). As the source is not an ideal point but has a finite size, the primary shadow (umbra) of the image is accompanied by a secondary shadow (**penumbra**) at the edges. This is also called the **Geometrical Unsharpness** $U_g$. By similar triangle method one can see that

$$U_g = ft/D_0 \qquad (5.3)$$

't' is the specimen thickness when the film is in tight contact with the specimen, but if there is a gap between specimen and the film, OFD (Object to Film Distance = t + gap) is used in place of t in Eq. (5.3).

$$U_g = \frac{f \times OFD}{D_0} \qquad (5.3a)$$

As Fig. 5.2 shows, the penumbra keeps enlarging as the film goes away from the job bottom. The $U_g$ can be minimized if

- The source is smaller in size
- $D_0$ is as large as possible

**Fig. 5.2** Geometrical unsharpness and magnification

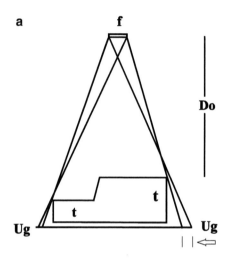

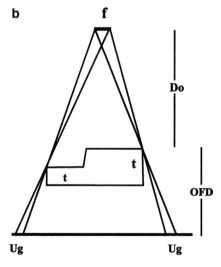

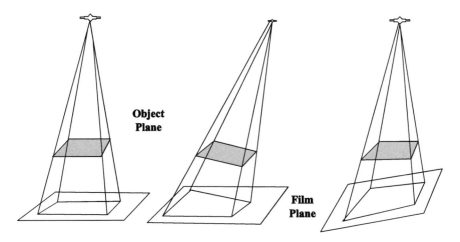

**Fig. 5.3** Images when the beam is not normal to the job or the film

- Film is in contact with the specimen bottom.

When film is away from the job, there is a magnification of the shadow. This direct enlargement is given as $M = = SFD/D_0$, i.e. $= SFD/SOD$.

Re-writing formula (5.3a) using this

$$U_g = f(SFD - SOD)/SOD \quad \text{or}$$
$$\mathbf{U_g = f(M - 1)} \tag{5.4}$$

Concept of direct enlargement is occasionally used in Fluoroscopy and mainly in Microfocal radiography (see Chap. 9).

### 5.2.1   Image Distortion

While taking a radiograph, the radiation beam should be perpendicular to the specimen and the film, or else the image will have distortion in such a way that an uneven magnification across the object will enter the image. See Fig. 5.3. The orientation affects defect interpretation as the flaw shape changes. It can even miss a defect.

## 5.3   Focal Spot Measurement

Measurement of the focal spot size is not a routine requirement for a field radiographer. Simplest method for this is the Pin Hole Camera Method. A block of high Z material like Pb or Tungsten with a fine hole in the centre is used for this. Focus area of the target is imaged through this hole onto a film with focus and film at equal distance from the block. Image of the focal spot on the film gives its size with some corrections. The hole has to be of the order of 1/10th dia as compared to the focal spot size. Hence the method is useful only for larger sizes of the focal spots. Focal spots on Minifocus (0.2 mm) and Microfocus ($< 100 \mu$) units need to be measured by different methods (see Chap. 9). Some of those methods (like use of Line Spread Function) can be used for large size focus also.

## 5.4   Scatter Control

As seen earlier, scatter is the unwanted radiation reaching the film, particularly on the area of interest. It is prominent at intermediate energies. There are four sources of scattered radiation.

(i)   From the specimen itself, from the volume being radiographed.

(ii)   From the specimen but areas *outside* the zone of interest (for example, from parent metal while testing welds).

Both these are called 'internal scatter'.

(iii)   From floor or walls behind the film, called 'backscatter'.

(iv)   From the sides, coming from walls or structures around the specimen.

Solution to the problem includes restricting the beam size to the minimum required.

(a)   Use of **diaphragm** on X-ray unit and **collimator** on gamma source. Adjustable diaphragm can bring the beam to the width just enough to cover the specimen.

(b)   **Filter**: This is a thin metal sheet placed in the path of the beam either close to X-ray unit or on the job. It filters lower energies from the continuous spectrum of X-rays. The method is more effective for cylindrical objects where undercutting happens at the edges. Too much filtering may harden the beam so much that can reduce the contrast. For high energy sources, filter is kept close to the job.

(c)   **Masks**: This prevents the radiation reaching the area outside the zone of interest, particularly during the RT of castings. For symmetric shapes, lead sheet can be used. For irregular shapes like local depression etc., moulding wax or putty mixed with barium sulphate can be filled. Fine size metal balls (called 'shots') can also be used. Some liquids like Carbon tetrachloride or lead compounds can be used in place of barium putty for very irregular shapes.

Both these are exhibited in Fig. 5.4a for cylindrical components. The putty prevents the central circular region of the casting getting over-exposed due to scatter from the ring shape around it. Lead sheet 1 does the same from the outside. If the entire casting is the zone of interest, then lead sheet (2) can be used in place of sheet (1).

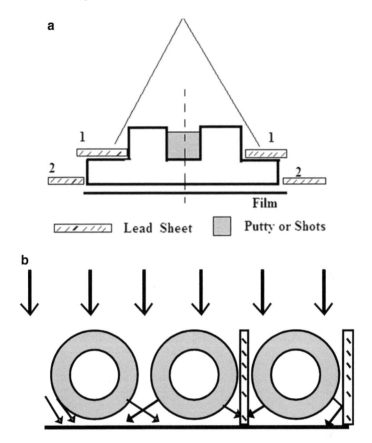

Fig. 5.4  **a** Use of masks, **b** prevention of undercutting

For back scatter, apart from back lead screen with the film, extra lead sheet can be used behind the cassette. In cylindrical objects tested simultaneously as in Fig. 5.4b, undercutting happens due to scattered photons going under the neighbouring component and cause overexposed edges. Vertical separator of lead or shape correction blocks can be used to prevent this.

(d) **Compensating Blocks**: These are used to reduce thickness variation along the beam path, like in cylindrical tubes or for fillet welds.

## 5.5  Double Film Technique

This technique can be used to cover wider thickness range in a single exposure. In this, two films are loaded in the same cassette, which can be of **same** speed or of two **different** speeds, one slower and the other faster. When both are of same speed, they will have identical image on them. Thin sections can be viewed on either of the films and two films stacked together are viewed for the thick sections, called **Composite Viewing**. Density is an additive function, thus if both films have lower density, say 1.2 D on the thick segment, the composite density will be 2.4 D, which is acceptable.

If films are of different speeds, faster film will register thicker sections of the casting and the slower one will register thin sections. One has to find the feasibility of this technique with the help of H & D Curve for the films (see Examples 8, 9 and 10 below).

**Nugget**

Some Standards, for Example IS: 1182 (RT of Butt welded Steel Plates) and some German Codes use a different approach to ensure a good geometry. Instead of directly limiting Ug value, it specifies a minimum Source to Object distance ($D_0$) in relation to the focal spot size. If the job thickness is 't' then it specifies

$$D_0/f \geq 7.5 \, t^{2/3} \text{ for normal technique}$$

$$D_0/f \geq 15 \, t^{2/3} \text{ for sensitive techniques}$$

Even in formula (5.3a) for Ug, ratio $f/D_0$ does occur. For a given focal spot size, $D_0$ has to increase with the job thickness in both the cases. But Ug is not specified here directly.

## 5.6  Numericals

*Example 1* A copper plate of 10 mm is to be radiographed using 400 kV X-rays. But the chart is available only for steel. For what thickness of steel the technique chart for steel should be referred to?

*Solution*: Use Table 5.1 for Equivalence Factors. Factor for Cu at 400 kV is 1.4, when steel is 1.0. Steel equivalent value of Cu is $10 \times \frac{1.4}{1.0} = 14$ mm.

The chart will be referred to for 14 mm steel.

(If there is a confusion whether to divide or multiply by the Equi Factor, a simple cross check is as follows: Copper being radiographically denser than steel, the equivalent value of steel thickness should come higher than the Cu-plate thickness.)

*Example 2* 1 ½ in. thick magnesium block is to be RT tested using X-rays upto 100 kV energy. Use the technique chart for aluminium (Fig. 5.1) and determine kV and exposure for density 2.

*Solution*: Table 5.1 indicates that Equivalence Factor for Magnesium w.r.t. Al is available and it is 0.6 for ALL energies.

Therefore Al equivalent of Mg is $1.5'' \times \frac{0.6}{1.0} = 0.9''$. The technique chart shows that 70 or 90 kV can be used for this thickness as the point falls on the mid portions of the lines. Select lower one i.e. 70 kV. (Two reasons for this. First, lower kV gives better contrast as we shall see in Chap. 6. Secondly, Mg having lower atomic number and density, bias should be for lower kV even after converting the thickness.)

Mark 0.9″ on the graph by a ruler scale; a line dawn towards 70 kV curve meets it in 24 mA min. This is the exposure required.

**Example 3** A welded Aluminium plate of 2″ thickness is to be radiographed at density 2.5 using the technique chart of Fig. 5.1. Find the exposure for suitable kV. The film H & D curve is of Fig. 4.2 (Chap. 4).

*Solution*: Suitable kV on the chart are 90–110 kV. Selecting the lowest value of 90 kV reads out the exposure to be 30 mA min. However the chart is made for density 2 and we need 3.0 D.

Log rel exp for 2.0 D = 2.7
Log rel exp for 3.0 D = 2.95 Difference = 0.25
Anti log 0.25 = 1.778.
(By calculator find $10^{0.2}$ this way: [10][$x^y$][0.25] result is 1.778.)
Multiply 30 mA min by 1.778 to get 53.37 or **53 mA min Exposure to get density 3.0**.

**Example 4** The job in Example 3 above has a space constraint and the SFD can not be higher than 50 cm. Determine exposure time in that situation. Tube current is 5 mA.

*Solution*: Technique chart is made at 70 cm, thus the answer found above related to 70 cm SFD. Apply modified Inverse Square Law Eq. (5.1) to find exposure at 50 cm.

$$E_2/E_1 = D_2^2/D_1^2 \quad \text{or} \quad E_2 = \frac{E_1 \times 50^2}{70^2}$$

where $E_1$ is 48 mA min as found above. Thus $E_2$ is 24.5 mA min. As the mA is 5, the time will be 24.5/5 = 4.9 min.

**Example 5** ASME code permits geometrical unsharpness of 0.76 mm for job thickness of 2″ (50 mm) and above. Whether the proposal to keep SFD of 50 cm is acceptable? X-ray focal spot is 1.5 mm.

As per Eq. (5.3), $U_g = f\, t/D_0$ where $D_0$ is SFD − t, f is 1.5 mm

$$U_g = \frac{1.5 \times 50}{500 - 50} = \frac{75}{450} = 0.167 \text{ mm which is } < 0.76 \text{ mm. Hence } \textbf{Acceptable}.$$

**Example 6** A lead plate of 0.75″ needs to be tested with Co-60 isotope at density 2.5 D. Use the Technique Chart for Steel for Co-60 given here (refer Fig. 5.5) and take

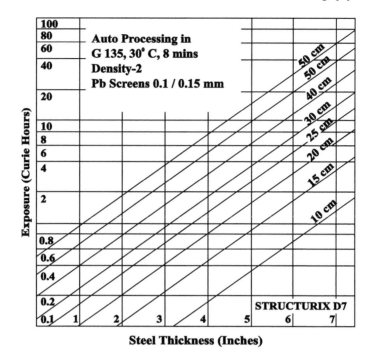

**Fig. 5.5**  Exposure chart for steel/Co-60

SFD 50 cm. Film used is 3 times slower than the one used in the Technique Chart and follows the characteristics of film C in Fig. 4.3 in Chap. 4. Find the exposure time for 4 Ci (148 GBq) source strength.

*Solution*: Referring to RT Equivalence Factors, one finds that lead (Pb) is 2.3 times more absorbing than steel if Co-60 is used. Thus the plate is equivalent to $2.3 \times 0.75''$ $= 1.725''$ of Steel. Next, refer to the chart in Fig. 5.5 for exposure for this thickness at SFD 50 cm. ($1.725''$ can be approximated to $1.75''$ for the purpose.) It is close to 1.2 Ci-hrs, but for the film D7. For the film C to be used, it will be $1.2 \times 3 = 3.6$ Ci hrs.

Now the correction for Density. Instead of 2.0 D on the chart we need 2.5 D. The H & D Curve (Film C) in Fig. 4.3 in Chap. 4 shows Relative Exposure values of 30 and 37 respectively for these densities.

As the x-axis is on Log scale, only ratio needs to be taken for finding new exposure (Anti-log is not required).

The Exposure should be $3.6 \times 37/30$ (because we expect it to increase, not decrease, for higher density).

This is $= 4.44$ Ci hrs. If the source activity is 4 Ci, the **time** of exposure will be **1.11 h**.

***Example 7*** What is the minimum SFD permissible for a 25 mm plate when radiographed with Ir-192 isotope with 3 mm source. Permitted $U_g$ is 0.5 mm. Consider two cases

1. If film is in contact with the specimen
2. When film is kept 1″ (25 mm) away.

*Solution*:

(1)  Use $U_g = f\,t/(SFD - t)$ you get $(SFD - t) = f\,t/U_g$

  i.e. $3 \times 25/0.5 = 150$.
  If $SFD - 25 = 150$, $SFD = 175$ mm. This is min permissible.

(2)  When film is 25 mm away, 't + gap' i.e. OFD becomes $25 + 25 = 50$ mm.

$$Ug = \frac{f \times OFD}{SFD - OFD} \quad \text{or} \quad SFD - OFD = \frac{f \times OFD}{Ug} = 150/0.5 = 300\,\text{mm}$$

  $SFD - 50 = 300\,\text{mm} \quad \text{or} \quad SFD = 350\,\text{mm}.$

(Note that the minimum SFD is higher when the film is not in contact.)

***Example 8*** A steel casting has two thicknesses viz 2″ and 1.25″. It is tested with Ir-192 using the technique chart given here (Fig. 5.6), on the film having H & D Curve of Fig. 4.2 (Chap. 4). If acceptable density range is 2–3.5 D, determine whether the casting can be radiographed in a single exposure on a single film?

*Solution*: On the exposure chart, select 24″ as SFD. This gives exposure of 12 Ci-hrs for 2″ and 5 Ci hrs for 1.25″ thickness. The ratio of exposures is thus 2.5.
  Now go to H & D Curve. If log rel values for given densities do not differ more than 2.5, exposure on same film is feasible.

Log rel exp for 3.5 D is 3.1 and log rel exp for 2.0 D is 2.85

$\Delta$ (log rel exposures) is $3.1 - 2.85 = 0.25$

anti log 0.25 is 1.778. This is less than 2.5, thus it is possible to accommodate both the thicknesses on the same film, single exposure (Fig. 5.6).
  Try the same example on a film with higher gradient. Perhaps it may not become possible to radiograph simultaneously or the margin of $\Delta$ (log rel exposures) may come down.

**Fig. 5.6** Exposure chart for
steel/Ir-192

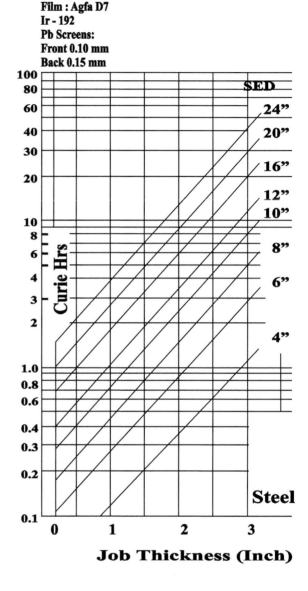

**Film : Agfa D7**
**Ir - 192**
**Pb Screens:**
**Front 0.10 mm**
**Back 0.15 mm**

*Example 9* Determine if a steel component with thicknesses 17.5 and 25 mm can be radiographed on the D7 film for which the technique chart is given here (Fig. 5.7). Allowable density range is 2.0–3.5. Use the suitable kV. H & D Curve of Fig. 4.2 in Chap. 4 to be used for this.

*Solution*: As the given thicknesses come at the extreme edge of the 160 kV line and 220 kV lines, only 180 and 200 kV to be considered. Exposures for these energies for 15 and 25 mm are

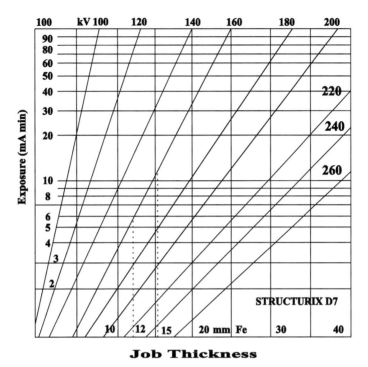

**Fig. 5.7** Technique chart for steel for X-rays

|  | 180 kV | 200 kV |
|---|---|---|
| 17.5 mm | 7.5 mA min | 4.2 mA min |
| 25 mm | 26 mA min | 13 mA min |
| Ratio | 3.4 | 3 |

For simultaneous RT, if 17.5 mm thick portion is close to higher density (3.5 D), then 25 mm should be above the lowest permissible (2.0 D).

Difference in Log Rel Exp for densities 2.0 and 3.5 is $(3.0 - 2.7) = 0.3$ (on H & D Curve of Fig. 4.2 in Chap. 4)

$$\text{Antilog } 0.3 \text{ i.e. } 10^{0.3} = 2.$$

That is the ratio of exposures to achieve these densities simultaneously. However the exposure ratios form the technique chart are 3 and 3.4 as seen above. Hence these two thicknesses can not be accommodated on the given film at 180 and 200 kV.

## Other approach

Let us work with 200 kV as an example. (Higher kV gives lower contrast and higher latitude, hence more likely to achieve the goal here.)

Let us examine whether 25 mm section reaches density 2.0 when 17.5 mm gets highest density i.e. 3.5 D. Steel thickness of 17.5 mm needs 4.2 mA min to get density 2. (Technique Chart is for 2.0 D.) By H & D Curve it needs to be doubled to get 3.5 D. (Can be seen from Fig. 4.2 in Chap. 4. Already seen above.) i.e. needs 8.4 mA min.

When you give 8.4 mA min to 25 mm thick section it will NOT reach 2.0 D because it needs 13 mA min as can be seen from the chart. Hence two section **can not** come on the same exposure.

***Example 10*** Check if the above specimen can be radiographed using Double Film technique on film II given in the Fig. 5.8 below.

*Solution*: We are told that both the charts belong to same film. (i.e. Film II is same as D7 of Fig. 5.7.)

Question here is about Composite Viewing of two films of SAME speed. Hence thin section (17.5 mm) can be taken to 3.5 D for single viewing and at that time 25 mm section should have at least 1.0 D for **composite** viewing, so that total of two films becomes 2.0 D.

Let us again use 200 kV.

Exposure for 17.5 mm thickness is 4.2 mA min. To take it to 3.5 D what exposure required?

Log rel exposure at 3.5 D is 2.8

Log rel exposure at 2.0 D is 2.52     Difference is 0.28

Antilog 0.28 is $= 10^{0.28} = 1.905$ (see log tables OR on calculator use keys $[10][x^y][0.28][=]$ to get reply)

So you need 4.2 mA min $\times$ 1.905 $= 8$ mA min.

**Fig. 5.8** Characteristic curve (for Numericals)

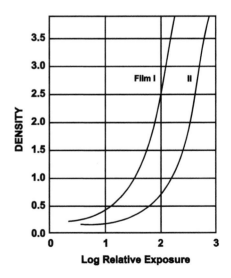

On the other hand 25 mm needs 13 mA min for 2.0 D. How much for 1.0 D?
Log rel exposure at 1.0 D is 2.2
Log rel exposure at 2.0 D is 2.52     Difference is 0.32
Antilog 0.32 is $= 10^{0.32} = 2.1$ times LESS exposure. So needs $13/2.1 =$ 6.22 mA min.

But we have given 8 mA min. Hence it will have density $> 1.0$ and **can be seen** in composite viewing.

**Example 11** Can the component of Example 8 be radiographed with double film technique, on different speed films? Take films I and II in above Curve (Fig. 5.8).

*Solution*: We have assumed that film in Fig. 5.7 is same as film II of Fig. 5.8. However we do not have Technique chart for film I. In this situation we work out relative speeds of I and II at a particular density. Say we find it at 2.0 D.

Using Fig. 5.8, log rel expo at 2.0 D for film I is about 1.85 and for film II it is 2.52. Difference is $2.52 - 1.85 = 0.67$

Antilog $0.67 = 10^{0.67} =$ on calculator $[10][x^y][0.67] = \textbf{4.67}$. This is **the speed ratio** of the two films.

25 mm has to be imaged on Faster film i.e. film I.
17.5 mm has to be imaged on slower film i.e. film II.

As two films are different, latitude is not the issue. Hence one can work with 180 kV also. Exposures for two thicknesses for 2.0 D on film II are

$$17.5 \text{ mm} \ldots\ldots 7 \text{ mA min}$$
$$25 \text{ mm} \ldots\ldots 28 \text{ mA min}$$

But 25 mm has to be taken on film I, which is 4.67 times faster. So the time required will be $28/4.67 = 6$ mA min. However we are giving 7 mA min anyway.

Thus thick section will have density $> 2$ and thus acceptable.

**Example 12** Find the exposure time for RT of 50 mm steel block using Co-60 source of 6 Ci (222 GBq) for SFD of your choice, but fulfilling the Ug limit of 0.5 mm. Film factor for the film used is 1.2 for 2.0 D. Source size 3 mm. HVT for steel for Cobalt 21 mm.

*Solution*: The formula for isotope source Eq. (5.2) to be used here.

But before that SFD to be selected.

$U_g = \frac{3 \times 50}{SOD}$ which gives SOD as 300 mm and minimum SFD $(= SOD + t)$ is 350.

Select SFD as 500 mm (0.5 m) and substitute in the above formula

$$\text{Exposure time (h)} = \frac{1.2 \times (0.5)^2 \times 2^{(50/21)}}{6 \,(\text{Ci}) \times 1.3 \,(\text{RHM})}$$
$$= \frac{1.2 \times 0.5^2}{6 \times 1.3} 2^{\frac{50}{21}}$$

$2^{\frac{50}{21}}$ can be found by calculator either directly by selecting following buttons

$$[2][x^y][([50][/][21][)][=] \quad \text{which gives } 5.2088$$

Or one can first find 50/21 (which is 2.38), then raise it on 2 by $[2][x^y][2.38][=]$ and get the same answer.

Continuing with the formula, Exp Time $= \frac{1.2 \times 0.25}{7.8} \times 5.2088 = 0.2\,\text{h or } 12\,\text{min}$. **Ans.**

### Model Questions

Q.1  For the Geometric Unsharpness to be minimum, one requires

   (a)  Smaller focal spot and small Source to Object distance
   (b)  Smaller Film to Object distance and larger focal spot
   (c)  Larger Source to Object distance and larger Object to Film distance
   (d)  Smaller focal spot and larger Source to Object distance

Q.2  A large source size can be compensated by

   (a)  Increasing Object to Film distance
   (b)  Increasing Source to Object distance
   (c)  Use of additional lead screens
   (d)  Lower exposure time

Q.3  By inserting the additional filters the average energy of the X-ray output

   (a)  Decreases
   (b)  Has no effect
   (c)  Increases
   (d)  Can not say

Answers are available in the section "Answers to the Model Questions".

# Chapter 6
# Quality of Radiographic Image

## Contents

## 6.1 Attributes of an Image

The properties defining a radiograph are

- **Optical Density**
- **Contrast**
- **Resolution** (also called **Definition** or Sharpness in radiography).

Density was defined in Eq. (4.1) in Chap. 4. As our eyes can not detect a small change in density $\Delta D$ at lower density, a minimum level of density is recommended in RT. For example ASME needs min 1.8 D for X-ray radiograph and 2.0 D for gamma rays.

**Contrast** is defined as the difference in density between two adjacent areas. Many factors contribute to contrast. Two main categories with subordinate factors are the subject contrast and Film contrast:

Subject Contrast

- Specimen's composition (atomic numbers), density, thickness

© Ind. Society for Non-Destructive Testing 2024
P. R. Vaidya, *Guidebook for Radiography*,
https://doi.org/10.1007/978-981-99-8038-3_6

- Energy used
- Scatter control including correct choice of Pb screens.

  Film Contrast

- Type of Film (mainly the Gradient of H & D Curve)
- Film Processing (type of developer, Time and Temp of development)
- Screen Type (Fluorescent screens have better contrast).

## 6.2  Definition

The conventional concept of resolution of an image is replaced by definition in radiography. Resolution is the ability to show two points separately; instead, here sharpness is more important. Thus definition is a **measure of sharpness** of the image. It is controlled by three factors:

Geometry Related

- Expressed by Geometrical Unsharpness $U_g$ (focal spot size, job thickness, SFD and job to film distance)
- Relative alignment of job and film with respect to source.

  Film Related

- Type of film and screens (emulsion thickness, grain size)
- Processing and graininess
- Film screen contact.

  Combined Factors

- Inherent unsharpness $U_i$ which depends upon radiation energy apart from film variety
- Movement Unsharpness caused by a relative movement between source, specimen and film during the exposure.

## 6.3  Latitude

Latitude is the range of specimen thickness which can be radiographed in a single exposure.

Refer to Fig. 6.1. These both are the radiographs of the same step-wedge. The upper figure has higher contrast but the radiograph is effective only for the 3 steps. In the lower image, entire wedge can be interpreted in the correct density range; it means high latitude in terms of thickness. But the contrast between steps is LOWER. If there is a requirement to have higher latitude, a film with lower gradient and/or higher radiation energy are employed.

In summary, when the contrast is higher, latitude is smaller.

**Fig. 6.1** Contrast versus
latitude

**HIGH CONTRAST - LOW LATITUDE**

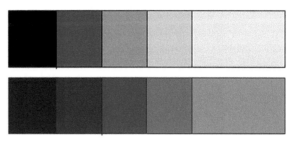

**LOW CONTRAST - HIGH LATITUDE**

When the contrast is low, latitude is higher.

## 6.4   Sensitivity

This is the combined measure of contrast and definition of a radiograph. The *contrast sensitivity* is measured as the per cent of the specimen thickness.

$$C\,(\%) = \frac{Smallest\ Variation\ detected\ in\ material\ Thickness}{Job\ thickness} \times 100$$

If $\Delta x$ is the smallest thickness change that can be detected in the specimen thickness of 'x', and $\Delta D$ is the smallest density change an eye can see,

$$\frac{\Delta x}{x} \times 100 = -2.3\Delta D/x\mu G_D \tag{6.1}$$

where $G_D$ is the film gradient at density D. However, a poor definition can influence contrast, particularly for features with narrow widths. Thus a concept of Image Quality Indicator (**IQI**) or **Penetrameter** comes. It is the tool to measure the *overall* sensitivity of the radiographic technique, combining contrast and definition. Usually the material of the IQI is same as that of the specimen being tested or is at least radiographically equivalent.

$$IQI\ Sensitivity\,(\%) = \frac{Thinnest\ element\ detected\ on\ the\ penetrameter \times 100}{Total\ Specimen\ Thickness} \tag{6.2}$$

However the **size of a penetrameter element seen on the film does not guarantee that the same size of flaw is being detected**. This is because the features on the IQI are regular shaped and the flaws are not.

## 6.5   Types of IQIs

Different Codes have different philosophy of determining the minimum dimension detected. This gives various designs of IQI.

1. *Step Hole type.* Used by the Indian Standard and the British Standard in rectangle shape. Thinner penetrameters are in hexagonal shape to for mechanical strength (Fig. 6.2). The French AFNOR uses triangular steps. The thinnest step on which the hole is seen is taken as the sensitivity achieved. Resolution and contrast are thus simultaneously assessed.
2. *Strip–Hole type.* Also called Plate–Hole type or Plaque. Design first used by ASTM and ASME, now adopted by many other Standards. While using these IQIs, a combination number $(X - nT)$ is quoted instead of only % sensitivity. X is the IQI thickness as % of the job and n shows which dia hole is seen; e.g. $2 - 4T$ etc. The US Military and Petroleum Industry Standards also use this design with different hole size combinations.
3. *Wire type.* Originally used by German Standard DIN, and adopted by ISO. Now used by European Standards EN. Later, the Indian Standards and ASTM have also adopted this design.

Benefit of wire (and step-hole) IQI is that one need not know *in advance* what sensitivity will be achieved; as the entire set of wire is placed, one which is seen decides the sensitivity.

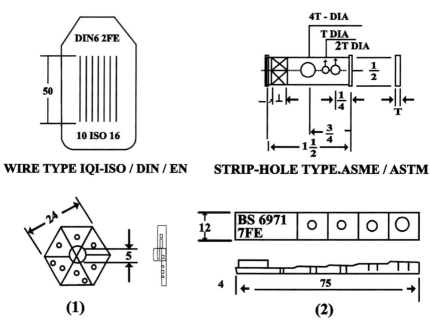

**Fig. 6.2**   Various types of IQIs

4. *Duplex Wire IQI*. It has 13 pairs of wires with different spacing. It is used especially for determining the total unsharpness. Thus wires are not from the material being tested but made of tungsten and platinum. Widely used in digital
   radiography to measure basic spatial resolution of the system.
5. *Others*. Japanese Standards JIS use wires as well as a stepped strip called
   'Contrastmeter' separately for contrast determination. Russian Standards GOST
   has grooves as the elements to detect.

As far as possible the IQIs are to be placed on the source side of the radiographic
set-up and on extreme edges of the field of view to make it a most 'unfavourable'
location. If they are kept on film side, it is easier to detect them because geometric
deterioration is nearly absent. Hence Codes specify thinner IQI if it *has* to be on the
film side. In case of weld testing, a shim is kept below IQI to match the density on
IQI and the weld (Table 6.1).

### 6.5.1  Equivalent Sensitivity (EPS)

This concept is defined by ASTM Standard but can be applied to any type of IQI or
arrangement where the contrast and resolution are independently measured.

$$\text{EPS} (\%) = \frac{100}{x} \sqrt{\frac{T\,h}{2}} \tag{6.3}$$

where x is specimen thickness, T is IQI thickness and h is hole dia, all in same units.

$$\text{It is also given as } n\sqrt{\frac{dia\ of\ smallest\ hole\ in\ T\ terms}{2T}} \tag{6.3a}$$

'n' is the IQI thickness as % of job thickness (Table 6.2).

This concept effectively shows that imaginary thickness of IQI on which 2T hole
could have been detected. In practice it permits use of thicker or thinner IQI than the
one recommended, provided the hole determined by this formula is visible. Table 283,
Article 2, ASME Sec V shows replacement IQIs in this manner.

### 6.5.2  Modulation Transfer Function (MTF)

Though contrast and resolution look independent, they are inter-related. For very fine
objects the contrast drops as the imaging system can not maintain fidelity. The concept
of MTF combines the two and is mainly for use in filmless (digital) radiography
applications. We will revisit the idea in Chap. 11 in detail.

**Table 6.1** Wire IQI designation, wire identity and diameter (mm)

(a) As per DIN/ISO/Indian Standard (set of 3)

| Wire No. | 1 ISO 7 | Wire No. | 6 ISO 12 | Wire No. | 10 ISO 16 |
|---|---|---|---|---|---|
| 1 | 3.2 | 6 | 1.0 | 10 | 0.4 |
| 2 | 2.5 | 7 | 0.8 | 11 | 0.32 |
| 3 | 2.0 | 8 | 0.63 | 12 | 0.25 |
| 4 | 1.6 | 9 | 0.5 | 13 | 0.2 |
| 5 | 1.25 | 10 | 0.4 | 14 | 0.16 |
| 6 | 1.0 | 11 | 0.32 | 15 | 0.13 |
| 7 | 0.8 | 12 | 0.25 | 16 | 0.1 |

(b) As per ASTM Standard (set of 4)

| Set A | | | Set B | | | Set C | | | Set D | | |
|---|---|---|---|---|---|---|---|---|---|---|---|
| Wire diameter | | Wire identity | Wire diameter | | Wire identity | Wire diameter | | Wire identity | Wire diameter | | Wire identity |
| in. | mm | | in. | Mm | | in. | mm | | in. | mm | |
| 0.0032 | 0.08 | 1 | 0.01 | 0.25 | 6 | 0.032 | 0.81 | 11 | 0.1 | 2.54 | 16 |
| 0.004 | 0.10 | 2 | 0.013 | 0.33 | 7 | 0.04 | 1.02 | 12 | 0.126 | 3.2 | 17 |
| 0.005 | 0.13 | 3 | 0.016 | 0.41 | 8 | 0.05 | 1.27 | 13 | 0.16 | 4.06 | 18 |
| 0.0063 | 0.16 | 4 | 0.02 | 0.51 | 9 | 0.063 | 1.6 | 14 | 0.2 | 5.08 | 19 |
| 0.008 | 0.20 | 5 | 0.025 | 0.63 | 10 | 0.08 | 2.03 | 15 | 0.25 | 6.35 | 20 |
| 0.01 | 0.25 | 6 | 0.032 | 0.81 | 11 | 0.10 | 2.54 | 16 | 0.32 | 8.13 | 21 |

**Table 6.2** Equivalent Sensitivity for different combinations

| Radiographic quality level | IQI as % of specimen thickness | Perceptible hole | EPS % |
|---|---|---|---|
| 2 − 1T | 2 | 1T | 1.4 |
| 2 − 2T | 2 | 2T | 2.0 |
| 2 − 4T | 2 | 4T | 2.8 |
| 1 − 1T | 1 | 1T | 0.7 |
| 1 − 2T | 1 | 2T | 1.0 |
| 4 − 2T | 4 | 2T | 4 |

## 6.6 Numericals

**Example 1** Wire No. 6 (1 mm) in DIN ISO penetrameter is visible on 40 mm thick specimen. What is the sensitivity achieved?

*Solution*: Using Eq. (6.2), WPS $= 100 \times$ wire dia/job thickness $= 100 \times 1/40 = 2.5\%$.

**Example 2** If Wire No. 5 from ASTM wire set A is seen well on the job with 14 mm thickness, what is the sensitivity achieved?

*Solution*: Wire No. 5 in Set A of ASTM wire IQI has diameter of 0.2 mm.

$$\text{WPS} = 100 \times 0.2/14 = 1.4\% \text{ sensitivity}$$

**Example 3** As per Table 276 for IQI Selection in ASME Section V, hole type penetrameter No. 15 can be placed on specimen thickness 0.25″ thro 0.375″. What is the range of sensitivity this provides, as approved by the Code?

*Solution*: ASME pene 15 is 0.015″ thick.

When it is placed on 0.25″ sensitivity indicated is $\frac{0.015 \times 100}{0.25} = 6\%$.

On the other hand, for the thicker end of the range (0.375″) it will give sensitivity of

$$\frac{0.015 \times 100}{0.375} = 4\%$$

Thus Table T 276 accepts sensitivity in the range 4–6%.

**Example 4** On a job of 30 mm thickness, ASME IQI no. 20 is seen with 4T hole. How will the sensitivity be described? What will be the Equivalent Sensitivity?

*Solution*: ASME 20 IQI thickness in mm is 0.5 mm. On 30 mm job it makes

$$\frac{0.5 \times 100}{30} = 1.7\%$$

By ASTM/ASME $(x - nT)$ convention this should be written as $(1.7 - 4T)$ sensitivity.

To find EPS let us use Eq. (6.3). $X = 30$, $T = 0.5$ and $h = 2$ (all mm)

$$\therefore \text{EPS} = \frac{100}{30} \sqrt{\frac{0.5 \times 2}{2}} = \frac{100}{30} \times 0.71 = 2.34\% \text{ Ans.}$$

(Here the contrast 1.7% is better than 2%, but the hole is 4T instead of 2T. EPS concept says if you need to see 2T hole, penetrameter of 2.3% i.e. thicker one only can show.)

## Model Questions

Q.1   A penetrameter is used to indicate

   (a)   The size of the discontinuity present in the test object
   (b)   The amount of film contrast
   (c)   The quality of radiographic technique
   (d)   Density of the film

Q.2   Definition of a radiograph is influenced by

   (a)   Source size
   (b)   Film graininess
   (c)   Geometry
   (d)   All the above

Q.3   What is true about the double film technique?

   (a)   Both the films should have the same speed
   (b)   Two films with differing H & D Curve can be employed
   (c)   It effectively increases the latitude of the radiography work
   (d)   All the above should be true

Answers are available in the section "Answers to the Model Questions".

# Chapter 7
# Radiography Application and Techniques

## Contents

The contents of this chapter will be useful also during Procedure Writing, as it will guide the technical decision for selecting the technique.

Radiography is used in following applications:

1. Quality Control of Welds
2. Quality Control of Castings
3. Quality Control of Forgings (occasionally)
4. To find internal details of assemblies (for various purposes).

## 7.1 Weld Testing

Welds come in a wide variety viz. plates, vessels, tubes, pipes, nozzles etc.

### 7.1.1 Plate Welds

This also includes Longitudinal welds (L-seams) on the large diameter vessels as they are as good as plate for RT purpose. Radiography procedure for a typical weld given in Fig. 7.1 will have following steps.

© Ind. Society for Non-Destructive Testing 2024
P. R. Vaidya, *Guidebook for Radiography*,
https://doi.org/10.1007/978-981-99-8038-3_7

**Fig. 7.1** RT of a plate weld

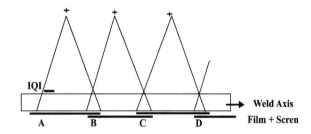

1. Decide the type of source to be used based on the material and thickness. If X-rays and isotope both are available/suitable and there is no site constraint, select X-rays.
2. Select the film: Slow films if very fine flaws are anticipated or high sensitivity is demanded or exposure times are too short to control. Otherwise medium film.
3. Decide the optimum parameters for exposure (kV, mA, SFD, time for X-rays; SFD and curie hours for gamma rays) for demanded density.
3a. For thick jobs, lower SFD will give sudden variation in density toward edges, longer SFD will give more uniform density.
3b. Verify that the SFD selected satisfies the Ug limits specified.
4. Decide the zone length. Although only a test exposure can decide exact zone length, an initial proposal is to be made in the procedure. As a rule of thumb, relation **SFD = 1.5 (zone length)** can be used for minimum SFD. This ensures that the density on the edges is not much lower than centre. Larger SFD is preferred. Large zone length may be acceptable in gamma exposures from density uniformity point of view. However larger source size for gamma ray may put a limit due to Ug needs.
    **Smaller zone is better as the orientation of a flaw remains favourable for detection.**
4a. Films should be larger than zone length so that the zone markers are seen in two adjacent exposures (see Fig. 7.1).
5. Select IQI as per the applicable Code or specification. A shim equal to weld crown height kept below the IQI, to keep density of weld similar to the IQI density.

### 7.1.2  Pipe Welds

A. *Longitudinal Welds:* These in large dia pipes can be treated as Plate welds as described above. Small diameter pipes where film can not be inserted inside need to be done in 'Double Wall Single Image' mode similar to circumferential welds. Penetrameter will be on the film side.
B. *Circumferential Welds*: Depending upon the diameter of the pipe, there are two broad approaches viz Single wall and Double wall RT. If radiation reaches the film after traversing only one wall, the flaw detection is better. This (called **Single**

**Wall Technique**) can be achieved either by placing the film inside and source on outer side or the other way round. However if geometry does not permit that, a double wall method has to be used. Figure 7.3 explains all options. Note the locations of the IQI in each. In addition note the following:

> Panoramic This can be used to save on exposure time. It can also be used when a vessel is closed but there is a hole to send Teleflex cable of isotope source inside. Calculate Ug taking radius of pipe as SFD. If it is more than permitted then the **source** can be placed in **off-set** position to increase SFD. In closed vessels, an entry point (like a nozzle) will be required for off set locations.
> DWSI **Double Wall Single Image technique** is needed for pipe diameter > 85 mm as per ASME; different Code can have different limit. This can be achieved by placing isotope directly on the pipe or from a long SFD; particularly in case of X-rays the latter is more convenient. When long SFD is used, an elliptical image is obtained, but the segment closer to the film is interpreted. A lead arrow is used to show direction of X-rays for identifying lower part of the weld. Number of exposures will be calculated from weld length with correct density in each exposure.
> DWDI **Double Wall Double Image** technique is for small diameter pipes (< 85 mm for ASME). The full diameter is exposed and the elliptical shaped image of the weld is obtained by keeping the source at an off-set angle. Two exposures at 90° are taken. One can also take three straight exposures without off-set, 120° apart. Top and bottom weld will be superimposed in this case.

## 7.1.3  Nozzle Welds

There are various designs of nozzle welds as per the relative sizes of nozzle vs shell and also with regard to configuration of setting of the nozzle on the shell. If hand can be inserted to fix a film, single wall RT is done here, or else go with double wall exposures. There are two faces of fusion line in the weld. The angle of shooting is selected such that at least one line of fusion is aligned with the radiation beam in order to detect a likely Lack of Fusion. Angle of 5°–7° is used for such cases. Sometimes 30°–40° is used to detect only volume defects when side wall LOF can not be detected due to geometry. In such cases ASME demands Ultrasonic Test also. (Ref ASME Sec III, NB-5243 for figures of such welds) The figure below gives some examples of shooting geometries for different types of nozzles and a Tee (Fig. 7.3).

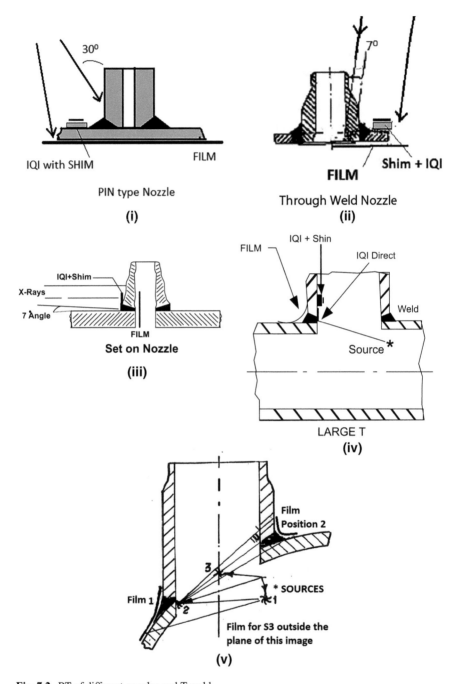

**Fig. 7.2** RT of different nozzles and T weld

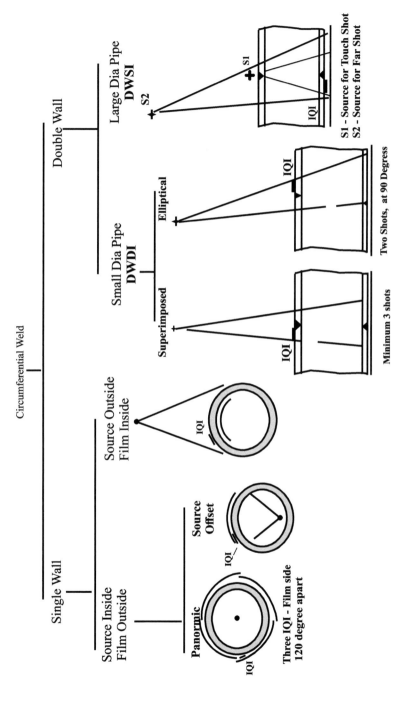

**Fig. 7.3** Flow diagram for pipe radiography options

**Fig. 7.4** Tangential RT set
up and image

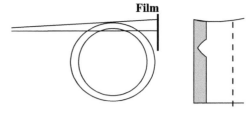

**Image with wall in profile**

## 7.2  Tangential Radiography

For certain applications radiation beam is aligned with the tangent to the pipe as shown in Fig. 7.4. The image will contain the profile of the wall; thus it is neither Single Wall nor Double Wall method. Hence a separate name for this technique.

It is mainly used for wall thickness measurement, thus frequently employed in measuring corrosion depth. This method is also useful for certain weld geometries. For example the pin type nozzle or set-on nozzle in Fig. 7.2 can be exposed by this technique by placing the film behind the nozzle on descending side of the shell. The weld will be seen in profile, somewhat like it is seen in the figure here. Any lack of fusion will be seen as dark line along the weld interface as can be seen on a longitudinal section. The technique is used in nuclear industry for the RT of fuel (see Chap. 12).

## 7.3  Casting Radiography

Very often the castings have intricate shapes and non-uniform thickness. Thus 100% RT is difficult but always attempted. Large number of exposures with small pieces of films inserted in different locations is the usual practice in the casting radiography (see Fig. 14.6 in the Annexure C of Chap. 14 on Procedure Writing). Use of double film technique also is useful mainly for RT of castings.

**Fillet Welds, Tube to Tube-sheet welds** and **spot welds** are generally exempted from radiographic examination.

**Nugget**

Relation between Zone Length and SFD

What should be the length of each zone in plate radiography? It is governed by two factors

(i) the angle of film plane with the radiation beam should not be so high that the defect shadows are distorted or missed.

(ii) the density towards the ends should not go below the acceptable value. The ratio of L and t in the figure depends upon the angle of radiation cone, which also controls ratio of zone length to SFD. (If L/t is 1.05, it means L is 5% longer than t.)

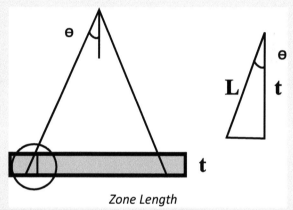

*Zone Length*

| $2\theta$ | L/t | Zone length |
|-----------|-----|-------------|
| $36°$ | 1.05 | 0.67SFD |
| $39°$ | 1.06 | 0.7SFD |
| $42°$ | 1.07 | 0.75SFD |
| $49°$ | 1.1 | 0.9SFD |

For better technique, lower angle is chosen. Whether it ensures the acceptable density range depends on latitude of the RT set up, dictated by source energy and film H & D Curve.

**Model Questions**

Q.1 A particular specification permits Double Wall Double Image RT of a 90 mm dia pipe (wall thickness 6 mm). If the source size is 2.5 mm and SFD is 400 mm, the Ug is

(a)  0.0375 mm
(b)  0.726 mm
(c)  0.0484 mm
(d)  0.5625 mm

Q.2  Usually in a panoramic exposure of a vessel or a large diameter pipe, the IQI placement rule is

(a)  one IQI per weld
(b)  at least 3 IQI over the circumference, equally spaced
(c)  one IQI per film
(d)  IQI on source side

Q.3  An acceptable quality radiograph should include

(a)  Zone and Location markers
(b)  Specified IQI with boundary and visible holes
(c)  Proper identification
(d)  All of the above.

Answers are available in the section "Answers to the Model Questions".

# Chapter 8
# Interpretation of Films

## Contents

## 8.1 Facilities Required

- A room with controlled lighting or a dark room
- Illuminator (also called Light Box) with adjustable aperture/diaphragm, variable light intensity
- Densitometer
- Magnifying Lens
- Optical comparator to measure flaw dimensions
- Reference radiographs when castings are to be evaluated.

### 8.1.1 Operator Attributes

- Good eye sight and acuity
- Requisite Certification
- Keen Observation
- Open Mind.

© Ind. Society for Non-Destructive Testing 2024
P. R. Vaidya, *Guidebook for Radiography*,
https://doi.org/10.1007/978-981-99-8038-3_8

He should have information about the weld joint configuration, weld process used. When seeing casting radiographs, he must know the casting process, shooting sketches.

**Dark Adaptation** is the first step towards interpretation. Eye functions by a different chemical process in darkness. Operator needs to be in dark for about 10 min to get best of the eye acuity in the dark. In lower light background, eye can see smaller difference in density shades ($\Delta D$). If he wants to visit outside the darkroom, he should wear goggles so that when he returns the adaptation is faster.

## 8.1.2  *What to Observe First?*

When the interpreter picks a film he will check the following:

- Does the film have proper optical density in the full zone of interest?
- Are identifications like Line No., Joint No., zone markings etc. are seen?
- Is the IQI designation (i.e. thickness or group no) as per the contracting document or specification?
- Is density difference between IQI and the weld (or area of interest in casting) within the permissible range? (e.g. ASME needs that the weld density be within + 30% and − 15% of the IQI density)
- Note any film imperfection (**artifact**) like scratch, water mark etc.

## 8.2  Vocabulary

**Indication**—A response to the probing entity (ultrasound, radiation or electromagnetic impulse) that requires interpretation to determine its significance.

**Flaw**—An imperfection in material or an item, which may or may not be harmful.

**Defect**—A discontinuity the size, shape, orientation of location of which makes it detrimental to the useful service of the part in which it occurs.

**Non relevant indications**—These are true indications produced due to discontinuities. However, the conditions causing them are present by design or accident, or other features of the part having no relation to the damaging flaws being sought. (Like punch marks, section thickness change etc.). The term signifies that such an indication has no relation to discontinuities that might constitute defects.

## 8.3  Weld Flaws

There are various reasons likely for each type of flaw that is seen. Hence the emphasis here is more on the nature and appearance of the flaw on the radiograph, than describing its cause. Other sections in your course may have that discussion. For easy reference, the flaws are listed in the alphabetical order and not as per the importance to the service.

**Burn-Through**—A local Collapse of weld pool. Seen as very dark broad spot usually surrounded by a ring of light density.

**Cracks**—Rupture of metal due to stress. Can be in the weld pool or at the boundary of heat affected zone with the parent metal. Sharp and zigzag lines, can be longitudinal or transverse. Hot cracks (called so, if it forms at temp above 200 °C) have branching. Cold crack are straighter and are found at Heat Affected Zone.

**Crater Crack** is a depression at the termination of a bead or in the weld pool beneath the electrode, contains a star shaped short crack.

**Diffraction Mottling**—A diffuse diffraction pattern (of light and dark lines or spots) on a radiograph resulting from X-ray in thin sections of material when grains have particular orientation.

**Icicles**—A coalescence of metal beyond the root of the weld. White round patches like frozen water droplets.

**Incomplete Penetration**—Root penetration which is less than complete or failure of a root pass and a backing pass to fuse with each other. On film, appears as elongated darkened lines of varying length and width which may occur in any part along the welding groove. If spacing is higher in weld preparation, this will be a dark band.

**Lack of Fusion**—Two dimensional defect due to lack of union between weld metal and parent metal (at root or at side wall) or between weld beads. A dark line, long or intermittent, near the weld groove. Side wall lack of fusion not easy to detect on radiograph, unless shot at an angle equal to bevel angle or has an accompanying cavity.

**Porosity**—Round shapes with defined boundaries; worm holes are cylindrical, thus dark spot if aligned with radiation. **Linear Porosity** is a straight line of fine pores, near central region.

**Root Cavity**—Dark areas of uneven shapes at the centre of weld; more likely in welds without insert rings and in carbon steel welds.

**Root Concavity**—See **Suckback**.

**Slag Inclusions**—Nonmetallic solid material entrapped in weld metal or between weld metal and base metal. On radiograph it could be seen as dark lines with some width; follows the shape of the bead when it is left out during chipping. Slag has density lesser than a cylindrical void of that size would have. Sometimes single point (of 'comma' shape) if an isolated piece has been trapped in molten pool.

**Slugging (Stubbing)**—The addition of a separate piece or pieces of material in a joint before or during welding. If not fused properly, same shape seen on film.

**Suck-Back (Root Concavity)**—Usually seen in pipe RT. Dark but diffused band at the centre, density uneven.

**Tungsten Inclusions**—Inclusions in welds resulting from particles or splinters of tungsten welding electrodes. On radiograph, a light area with rounded edges or filamentary shape. Very bright for aluminium welds; less so for Steel welds.

**Undercut**—A depression or groove adjoining the toe (i.e. outer edge) of the weld. Appears on a radiograph as a dark line or area of irregular shape, continuous or intermittent. Sometimes it comes on the sides of the ID bead also.

**Undercutting (Edge Burn Out)**—The excess blackening of the film within the image of any unblocked edge due to radiation scattered, or passing around, not through, the object. Seen on cylindrical and round objects.

**Wagon Tracks**—Comes due to voids or gaps at the root when a backing strip is used. Two parallel dark lines at the centre of weld with variation in width. If straight in some segment, may indicate LoF.

**Mismatch**—In pipe RT, mismatch of pipe faces gives sharp step of density mismatch, called 'high-low'.

There are reference radiographs available for welds from ASTM and IIW for training purpose. They are not used as a bench mark for acceptance/rejection. That decision will be as per the Acceptance Standard for the job.

## 8.4  Casting Flaws

Unlike welds, a flaw in a casting can occur in any part of the component volume. Also, it is not possible to assess depth extension of a flaw by radiograpy; hence acceptance standard here is usually by comparison with the reference radiograph set. Prominent among them are from ASTM, for variety of materials and thickness ranges. (Some standards have found methods to assign numbers for quality and acceptance.) Following is the description of appearance of casting flaws.

**Gas Holes**—Holes created by a gas escaping from the molten metal appear as round or elongated, smooth-edged dark spots, occurring individually, or in clusters, or distributed throughout a casting. (Usually voids with diameter > 1/16″ are called Blow Holes and smaller as Porosity.)

**Gas Porosity**—Refers to porous sections in metal that appear as round or elongated dark spots corresponding to minute voids visually distributed through the entire castings.

**Inclusions**—Usually sand inclusion when sand mould is used. Oxidized metal inclusion occurs in Copper and Al–Mg castings. (In Cu/Cu–Ni castings it is called **Dross**) Density lighter than a gas hole of same size. Sand inclusions have angular edges.

**Shrinkage Cavities**—cavities in castings caused by lack of sufficient molten metal as the casting cools, consists of voids which have different forms as per the cause and geometry of casting. Viz.

Conical voids (in cylindrical objects), spongy, feathery, dendritic.

*Appearance on radiograph*—Large dark voids easily identified. Other varieties have dark area of irregular shape with a dendritic, filamentary, or jagged appearance on minor magnification.

**Microshrinkage**—Fine shrinkage cavities detectable only at moderate magnifications (about 10×), consisting of interdendritic voids. This defect results from contraction during solidification where there is not an adequate opportunity to supply filler material to compensate for shrinkage. Alloys with a wide range in solidification temperature are particularly susceptible.

*Appearance on radiograph*: appear as dark feathery streaks, or irregular spongy patches that indicate cavities in the grain boundaries. When fine structure is not seen, the area is darker than the good portion.

**Hot Cracks**—Appear as ragged dark lines of variable width and numerous branches. They have no definite line of continuity and may exist in groups. They may originate internally or at the surface.

**Cold Cracks**—Appear as a straight line usually continuous throughout its length and generally exist singly. These cracks start at the surface.

**Hot Tear**—A fracture formed during the solidification process because of hindered contraction. Though this is also a crack and only the origin is different, this will have multiply woven cracks making it a broader one than the hot crack.

**Unfused Chaplets**—These tiny components are inserted in the mould to support the core. They must fuse after pouring. If they do not, their shapes are seen on films.

**Core Shift**—Asymmetry in the casting seen on radiograph can indicate if the core has moved during the process. For example, for a hollow cylinder, wall thickness will be different on two sides.

**Mottling**—A diffuse diffraction pattern resulting from X-rays in thick sections of crystalline material. Shows as Streaks of dark and light bands alternatively.

**Cold Shut**—A discontinuity that appears on the surface of cast metal as a result of two streams of liquid meeting and failing to unite. Can be seen on film as a straight line, sharp on one side; seen only when aligns suitably.

**Misrun**—When molten metal does not fill the mould completely a dark area/patches will be seen on the film. Can be seen on visual inspection also.

## 8.5   Forging Defects

**Burst**—Rupture of metal when the forging ratio is higher or temperatures are not commensurate. Dark area with irregular branching, looking like shrinkage in castings.

**Cold Shut**—a portion of the surface of a forging that is separated, in part, from the main body of metal by oxide.

## 8.6   Artifacts

When an indication on the film does not match with any likely flaw in the casting or the weld, one suspects that it has come from an Mass Attenuation.

| Indication | Reason |
| --- | --- |
| Dark spots | Splashing of developer solution on film BEFORE exposure |
| White/light spots | Splashing of fixer solution on film BEFORE exposure |
| Dark lines | Scratches on Lead Screens (or on the film: can be seen in reflection |
| White hair lines or other shapes | Hair or paper or film pieces/dirt between film and screen |
| Light spots (dots) | Air bubbles sticking to film when entering the developer |
| White crescent (crimp/nail) mark | Kinking of film after processing. (Or anytime, if emulsion is seen to have come off) |
| Dark crescent (crimp/nail) mark | Kinking of film before development |
| Yellow stains | Inadequate rinsing, inter-mixing of processing solutions |
| Milky appearance | Incomplete Fixing |
| Dark (black) streaks | Light passage through the Cassette or film exposed during handling |
| Star shaped dark sharp lines | Static electricity marks—during film handling |

(continued)

(continued)

| Indication | Reason |
|---|---|
| Dark and light streaks on the edges | Old residual solution in the hanger or frequent pulling out of film during developing or lack of agitation |
| Frilling (detaching emulsion) | Fixing or washing at high temperature |
| Reticulation | Large temperature difference between processing baths |
| *Artifacts related to auto processing*: | |
| Dark lines at regular intervals (Pi lines) | Fine deposits on the roller, transferred to film every rotation |
| Fine lines parallel to travel direction | Tight rollers |
| Dark spots with tails in travel direction | Dirt on film while entering the processor |
| Streaks | Processor used at long intervals or high developer temperature |

**Nugget**

When Hot tear is broad, it can resemble a linear shrinkage void. However.

Hot tear propagates near or at surface Shrinkage is at mid section by nature.

Hot tear comes when temp gradient is high Shrinkage comes if temp gradient is LOW.

Hot tear occurs where section change is there Usually Shrinkage also at section change; but can occur midsection.

Hot tear is in direction transverse to great stress.

(Extract from ASTM E–272 Reference Radiographs for High Strength Copper base and Cu–Ni castings).

**Model Questions**

Q.1  Separation of emulsion from the base is called

  (a)  Frilling
  (b)  Reticulation
  (c)  Splitting
  (d)  Puckering

Q.2  In a radiograph done at 120 kV, a light fine image of the shape of a human hair is seen. It could be due to

  (a)  An electrostatic mark
  (b)  A hair trapped between the specimen and the cassette
  (c)  A hair trapped between the lead screen and the film
  (d)  A hair lying on the top of the specimen

Q.3  In a steel weld, a jagged sharp line having occasional branches, could be

    (a)  a longitudinal crack
    (b)  a crater crack
    (c)  a transverse crack
    (d)  any one of the above

Q.4  Which of the following welding discontinuities would be most difficult to Image radiographically:

    (a)  Planar lack of fusion
    (b)  Incomplete penetration
    (c)  Undercut
    (d)  Slag inclusions.

Answers are available in the section "Answers to the Model Questions".

# Chapter 9
# Advance Techniques

## Contents

## 9.1 High Energy Radiography

Radiograpy at energy 1 MeV and above is considered to be 'High Energy Radiography'. Thus it includes Co 60 isotope source as well as bremsstrahlung producing machines viz.

- Resonant Transformer
- Van de Graaff Generator
- Linear Accelerator
- Betatron

© Ind. Society for Non-Destructive Testing 2024
P. R. Vaidya, *Guidebook for Radiography*,
https://doi.org/10.1007/978-981-99-8038-3_9

The last three are the particle Accelerators and operate on different principles than normal X-ray units. Except this, there is no major difference in actual practice of radiography *per se* at high energy. As for Co-60 isotope, if thick objects like Rocket Motors or Missile bodies are to be tested the source strengths of 100 to 1000 Ci (or TBq) are employed.

### 9.1.1  Resonant Transformer

Not popular anymore. Design concept is same as conventional X-ray unit except that the natural frequencies of primary and secondary windings are tuned with each other. To reduce the size of the coils, the operational frequency is higher than the line frequency (50 Hz). X-ray tube is designed to apply voltage in segments.

Energy: 1–2 MeV Focal spot: 7 mm Dose at 1 m: 50 rad/min
Tube Head Weight: 1000 kg Max Steel Thickness RT: 12 cm

### 9.1.2  Van de Graaff Generator

*Principle*: Accumulation of electrostatic charges to produce high voltage.

In Fig. 9.1, a continuously moving belt is sprayed with electrostatic charges at the bottom point and the same are collected by a comb above and deposited on to a metallic dome. This charge Q and the Capacitance C between two hemispheres causes a voltage to develop on the dome as per formula $Q = CV$. The high voltage is connected to the cathode of the accelerator tube to drive the electrons towards the target just as in normal X-ray tubes.

Energy: 1–3 MeV Focal spot: Up to 2 mm Dose at 1 m: 350 rad/min
Tube Head Weight: 3000 kg Max Steel Thickness RT: 30 cm

### 9.1.3  Linear Accelerator

*Principle*: Electron accelerated riding on high frequency electro-magnetic waves (microwaves).

Basic concept can be understood with the help of Fig. 9.2a. Here alternative cylinders are connected with the same polarity. The alternating field changes the polarity exactly when the particle exits one cylinder and enters the next. Hence it is attracted and thus accelerated. As the increased energy makes it faster, subsequent cylinders are longer in length. This concept is used for the heavy particles; but for electrons (with speeds close to the speed of light) the lengths of cylinders will be enormous.

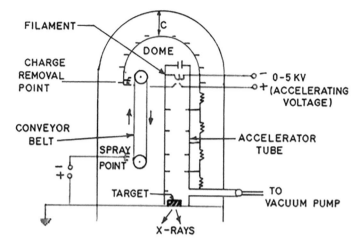

**Fig. 9.1** Schematic of the Van de Graff Generator

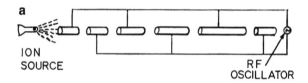

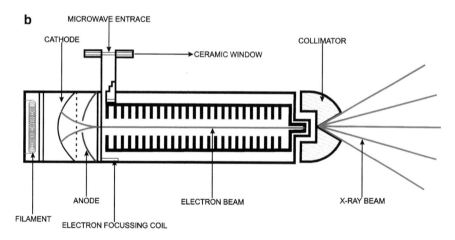

**Fig. 9.2** **a** Linear accelerator for heavy ions **b** the construction of a LINAC

Therefore Linear Accelerators for electrons use microwaves (frequency in GHz range) to accelerate and carry electrons. Microwave cavities of about a metre length are adequate, in which electrons travel. They are made to impinge on a target at the end of travel, where they produce bremsstrahlung radiation. Linear Accelerators (also called LINAC) are of two types, Standing Wave and Travelling Wave type; the latter is used for electron LINACs. They come in a wide range of energies and have very high radiation output.

Energy: 2–25 MeV Focal spot: 2–3 mm

Tube Head Weight: 1000–3000 kg Max Steel Thickness RT: 12–60 cm

Dose at 1 m:

200 rad/min for 2 MeV
1000 rad/min for 6 MeV
5000 rad/min for 15 MeV
10,000 rad/min for 25 MeV

### 9.1.4  Betatron

*Principle*: Electrons go in circular orbit in a magnetic field and their velocity increases with the field strength.

Betatron has two main components—a 'doughnut' shaped tube (or cavity) in which electrons go round for acceleration and the electromagnets to create magnetic field. The tube with vacuum inside is placed between two magnetic pole pieces and injected with electrons. After the required velocity is achieved the electron bunch is ejected out using a special coil. It then falls on a fine wire shaped target to produce X-rays.

Energy: 15–30 MeV (in RT) Focal spot: 0.2 mm Dose at 1 m: 100 rad/min

Tube Head Weight: 1500k Max Steel Thickness RT:50 cm

### 9.1.5  Features of High Energy RT

1. All the three accelerators give pulsed output.
2. All have transmission type target (in the conventional X-ray tubes the radiation beam comes out on the same side as electron beam, but here the X-ray beam is on the other side of the target as can be seen in the figures).
3. Radiation beam is narrow for high energies. The width reduces with increase in energy. Hence the Beam Flattening discs are used to broaden the beam to cover the large size objects within the radiation field.
4. Efficiency of X-ray production is higher (about 10%) than X-ray units.
5. At higher energies, scattered to direct intensity ratio is smaller.

6. Apart from Lead screens, here other metallic intensifying screens like Cu or S.S. are also used.
7. Inherent Unsharpness of films is higher at these energies. (e.g. 0.6 mm at 10 MeV).
8. Dependence of Mass Attenuation Coefficient ($\mu_m$) on atomic number reduces at high energies. (It is nearly constant at about $0.06$ cm$^2$/g). Thus the linear attenuation coefficient $\mu_L$ strongly depends upon density of material alone.

## 9.2 Microfocal Radiography

Conventional X-ray units have focal spot size ranging from 0.4 mm to 4 mm. This is what one can achieve with the filament of reasonably thick wire to last lifelong and a focusing cup around the filament to guide the electrons towards the target. Microfocal (MF) X-ray units use additional electromagnetic focusing coils around the electron path to bring it them to a fine focus size. The units are usually pump-down type constructions, which permits opening of the tube head and change the filament if it breaks. Thus a finer wire filament is possible to be used. Thus MF units are available with focal spots from 10 to 100 $\mu$m. New units—nanofocus units—have come to the market with focal spots of the order of 500 nm, with transmission targets (Fig. 9.3).

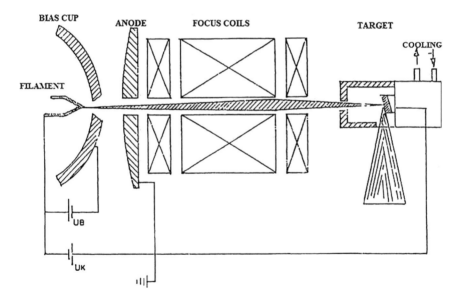

**Fig. 9.3** A schematic of a typical tube head in a microfocal unit

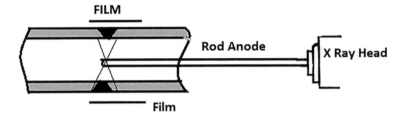

**Fig. 9.4** Panoramic radiography with rod anode

## 9.2.1  Benefits of MF Radiography

Refer to Eq. (5.3a) (Chap. 5) for geometric unsharpness with a gap between specimen and the film

$$U_g = \frac{f \times \text{Object to Film Distance}}{\text{Source to Object Distance}}$$

For MF unit the 'f' is very small. Hence,

(i)  the $U_g$ is proportionately smaller and the fine defects which would have got blurred out will now be detected.

(ii)  OFD (= t + gap) can be increased. As t is fixed, gap can be increased resulting in a magnification of the image by a factor M = SFD/SOD. M as large as 20 is common. This is not possible with conventional units. Magnification reveals fine flaws and makes interpretation easy. A gap between object and the film also helps in radiography of radioactive objects (refer Chap. 12)

(iii)  SOD can be reduced to a great extent without sacrificing sensitivity. This facilitates panoramic RT of small dia pipes/tubes. SOD is equal to the radius of the pipe. A special rod anode is employed for this. Figure 9.4 Rod anode can reach some inaccessible geometries also.

## 9.2.2  Focal Spot Measurement

This is more complex for MF units as compared to conventional units (see Chap. 5) The simplest method is to use magnified image of a wire and measure its $U_g$. The formula $U_g = f(M - 1)$ gives value of 'f'. More sophisticated methods use Resolution Patterns of different shapes in lead or gold alloys. Indirectly these methods use the principle that the edge spread function (ESF) and line spread function (LSF) of an imaging system depend upon the focal spot size as one factor. A derivative of LSF is Modulation Transfer Function (MTF) and that too is used.

## 9.3  Neutron Radiography

Neutrons can be used for RT because, just as photons they are also partly transmitted and partly attenuated (absorbed + scattered) within the specimen. In some way Neutron radiography (NRT) is complimentary to X-radiography (XRT) because.

(i)   Neutrons easily penetrate thick sections or high Z materials, for which XRT needs high energy source

(ii)  Neutrons have higher attenuation in hydrogenous and low Z materials. These are fairly transparent for XRT unless very low kV is used

(iii) If a component of low Z material is covered by high Z material, XRT can not show it; neutrons are particularly better for such a combination.

But it is different from XRT in one way. The X-ray films are not sensitive to neutrons as these chargeless particles do not interact with the electrons of the emulsion molecules. One needs help from **Converter Screens** to get an image on the films. These screens get activated by neutron bombardment a radiation (gamma or electron) which affect the film in proportion to the number of neutrons received by the screen. This in turn depends inversely upon the soundness of material, as in XRT. More radiation will pass through voids and thus give dark image. There are two types of converters.

(a)  those with a prompt radiation output like gadolinium

$$^{155}Gd(n, \Upsilon)^{156}Gd, ^{157}Gd(n, \Upsilon)^{158}Gd$$

(b)  those which emit radiation with longer half life like gold, dysprosium and indium

$$^{197}Au(n, \Upsilon)^{198}Au \quad \text{half life 2.7 days}$$

$$^{164}Dy(n, \Upsilon)^{165}Dy \quad \text{half life 140 min.}$$

$$^{115}In(n)^{116m}In \quad \text{half life 54 min}$$

$$^{115}In(n, \Upsilon)^{116}In \quad \text{half life 14 s}$$

### 9.3.1  Sources for NRT

NRT is carried out using Thermal neutrons because they are the ones which are prominently absorbed. But the sources mentioned below give fast neutrons. Their velocity is brought down by passing them through the hydrogenous materials (like water, paraffin etc.) or materials with low Z. This process is called '**moderation**' and brings these neutrons to thermal energies.

Nuclear Reactor

Generators using D –T reactions (deuterium-tritium)

Nuclear reaction based sources (Antimony Beryllium, Americium Beryllium)

Accelerator based where Be disc is placed after the target to give ($\Upsilon$, n) reaction

Spontaneous fission source Cf $^{252}$.

Just as beam intensity decides the exposure in XRT, flux of the source decides exposure time in NRT. Flux is given as no. of neutrons/cm$^2$/s. About $10^9$ thermal neutrons/cm$^2$ gives a radiograph. As there is no single point from where the neutrons come, the idea of focal spot is not there. Instead a parameter **L/d** is used to assess resolution capability of the set up; L and d are respectively the length and diameter of the collimator used. Higher the L/d ratio, better is the image.

### 9.3.2  Techniques

*Direct Technique*: Arrangement is similar to XRT. Converter screens of Gd are used (Fig. 9.5).

*Transfer Technique*: Often the neutron source may contain gamma rays along with neutrons which can affect the film kept behind the converter. Or the object itself is radioactive, like irradiated nuclear fuel, which could expose the X-ray film. In such cases only the delayed emission type converter screens are used WITHOUT the X-ray film. Once the screen gets activated it is taken to the dark room and placed in contact with film where it can keep exposing till desired time (Fig. 9.6).

**Fig. 9.5**  Direct technique

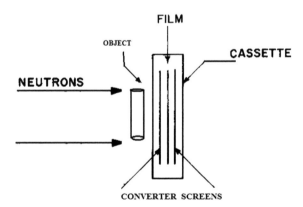

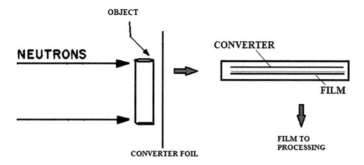

**Fig. 9.6**  Transfer technique

### 9.3.3  Track Etch Technique

This avoids the use of a film and a converter. A sheet of cellulose nitrate with a thin coating of lithium borate receives the neutrons and emits alpha particles by (n, α) reaction. This creates small tracks on the nitrate film which is further processed by placing in hot NaOH solution to make them deeper. That creates a profile of an image, which is here a radiographic image. The scheme is insensitive to gamma rays and thus is equivalent to Transfer technique.

### 9.3.4  Real Time Imaging

Like real time X-ray radiography NRT can also be done on Fluoroscopy screens, Image Plate etc. For the phosphor to be sensitive to neutrons it is doped with absorbers like lithium or boron. Other features are similar to X-ray RTR.

### 9.3.5  Applications

1. To find internal details in assemblies with light materials like nylon, teflon, plastic etc.
2. In specific cases where light material is expected to be the contaminant or impurity, for example hydrides in zircalloy components.
3. Inspection of artillery shells where the 'fuel' is low Z material (Fig. 9.7 note that the explosive charge being low Z material is seen with more contrast in the neutron radiograph whereas the bullet in high density material is better in X ray image [1]).

**Fig. 9.7**  Empty and filled bullets: Top—X-ray image, bottom—neutron image

## 9.4  Flash Radiography

If one radiographs an object which is in fast motion, linear or rotational, it will show movement unsharpness. However if it is exposed for extremely short time, this blur does not occur. This can be done by pulsed exposure where the pulse has very high radiation intensity. Though LINAC or Betatron are also pulsed sources, they can not deliver instantaneously. A special process called Flash Radiography is developed for this, where the tube current in a single pulse is of the order of **kiloamperes.** Typical pulse duration is 100 ns, so that even an object with velocity of 10 km/s will not move more than 1 mm during the exposure.

To achieve such quick emission, a process called '**cold emission**' (or **Field Emission**) is used instead of customary thermionic emission from a heated filament. A circuit with a bank of capacitors provides a momentary high voltage positive pulse which 'pulls' electrons out from the cathode surface and sends to the anode. This becomes the tube current. Apart from high intensity electron beams, gas discharge tubes and vacuum discharge tubes can be used as the sources.

### 9.4.1  Applications

- To study the movement of a ballistic covered by a smoke
- Observe rotating turbine or other machinery
- To study the burning process of a detonator charge
- To radiograph radioactive object, because this can reduce the time of contact of radioactive object with the film.

**Model Questions**

Q.1 In a microfocus X-ray unit, the small focal spot is created by

(a) Focussing the X-ray beam before target
(b) Placing a collimator on the X-ray tube
(c) Focussing the electrons going to the target
(d) None of these

Q.2 This is not the suitable detection medium for neutron radiography

(a) Lithium doped Image Phosphor Plate
(b) X-ray film with Pb screen
(c) X-ray film with Gadolinium screen
(d) Track etching film

Q.3 Van-de-Graff generator is used for radiography in order to

(a) Have a radiograph of a speeding object
(b) Get a radiograph of a radioactive object
(c) Have a higher energy radiation compared to conventional X-ray units
(d) Get an intense source.

Answers are available in the section "Answers to the Model Questions".

# Reference

1. Ghosh JK, Panakkal JP, Chandrasekharan KN (2005) Neutron radiography: a complementary technique to photon radiography. NAARRI Bull XII(1)

# Chapter 10
# Special Applications and Techniques

## Contents

These techniques are actually a collection of ideas that can be applied in particular needs but rarely used in industrial practice. Some of them have only text book significance.

## 10.1 Electron Radiography

This can be applied to very thin material sections or to get surface profile of the materials to be tested. An easy way to get electrons is to use photo electrons. There are two variations.

(i) *Transmission electron radiography*

This is used for RT of thin section of materials through which electrons can pass through; obviously they are non-metallic like paper, plastic etc. High energy X-ray beam generates photoelectrons from a sheet of lead (or any high Z material)

© Ind. Society for Non-Destructive Testing 2024

P. R. Vaidya, *Guidebook for Radiography*,

https://doi.org/10.1007/978-981-99-8038-3_10

**Fig. 10.1** **a** Transmission
method. **b** Emission method

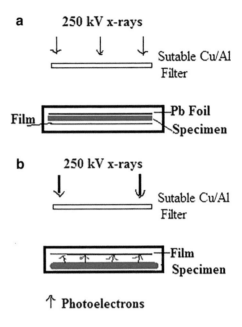

which is in tight contact with the specimen and a film behind that (see Fig. 10.1a).
The cassette can be evacuated to create a tight contact.

(ii)  *Emission electron radiography*

Photoelectrons here are produced from the surface of the specimen itself, which
has to be a metallic material. For better contact with the film, the specimen
surface is in polished condition (Fig. 10.1b).

## 10.2  Micro Radiography

After the advent of micro-focal X-rays, utility of this technique has become limited.
Essential part here is the very fine grain emulsion, so that the image can be magnified
on microscope for detailed study. Low kV X-rays need to be used to avoid film
inherent unsharpness; thus the specimen thickness is restricted. Local variations in
phases, segregation of alloying elements, fine discontinuity in solders/welds etc. can
be seen if their atomic numbers are widely different. It is also used for radiography
of biological tissue sections.

## 10.3   Depth of Flaw Estimate

(i)  **Stereo-radiography**

Two radiographs are taken for an object by shifting the X-ray unit by distance equal to the distance between eyes. These films are viewed in an arrangement where the left eye sees the left side radiograph and the right eye sees the right side radiograph but simultaneously. This gives a virtual 3-D image created by the brain in the mind, but it is a subjective feeling; the depth of flow cannot be *measured* by this method.

(ii)  **Parallax method (or triangulation method)**

Here also two exposures are taken by shifting the X-ray source but the shift is much more than the above (see Fig. 10.2). both the exposures are taken on the SAME film.

$\Delta S$ is the shift in the source position
$\Delta F$ is the shift in the image of the flaw
d is unknown and includes gap between film and object, if any.

**Fig. 10.2**  Parallax method

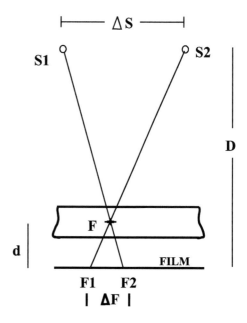

From similar triangles S1–S2–F and F–F1–F2, one can write $\Delta S/\Delta F = D/d - 1$.

Or

$$d = \frac{D\Delta F}{\Delta S + \Delta F} \tag{10.1}$$

As the displacement of the flaw image is not very large, accuracies are only moderate.

This method is also called the **Rigid Formula method**.

(iii) **Marker methods**

(a) *Single Marker*—To improve upon Parallax method, a marker in the form of a lead letter or wire is kept on the source side of the specimen. A shift in its image will be more than the shift in defect image. Assuming a linear relationship between these two shifts one can estimate flaw location as a fraction of specimen thickness. No need to measure source movement.

(b) *Double Marker*—A more accurate estimate can be obtained if one marker is placed on the top of the specimen and one on the bottom (Fig. 10.3). Here the film is kept away from the specimen to get even larger accuracy.

The shift in two markers and the flaw is given respectively by $\Delta$TM, $\Delta$F, $\Delta$BM. Location of the flaw at height 'h' from bottom can be given as

**Fig. 10.3** Double marker method F = Flaw, TM = Top Marker, BM = Bottom Marker

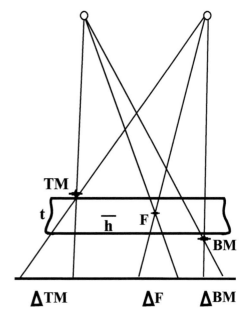

$$h = t \times \frac{(\Delta f - \Delta BM)}{\Delta TM - \Delta BM} \qquad (10.2)$$

## 10.4 In-Motion Radiography

In this technique the job and the film move together in a linear (or rotational) motion while receiving the radiation from the source, which is stationary. A narrow collimator slit provided above the specimen works as a mask also (Fig. 10.4). Thus exposure is continuous but only a small section is radiographed at a time,

Advantages

1. There is no scattered radiation reaching the film through the portions of specimen not being radiographed, as they are behind the collimator. This improves the image quality.
2. Time spent on exposure arrangement is eliminated.
3. No angular distortion of the flaws.
4. Uniform density (phenomenon of lower density on the extreme edges does not occur).

Limitations

1. As the exposure is very short, thick section specimen cannot be tested.
2. One needs Roll films, preferably loaded with lead screens.
3. Small amount of movement unsharpness occurs as the job travels from one edge of the slit to the other.

Applications: Plate type nuclear fuels (linear motion), locate motors (rotational motion).

**Fig. 10.4** In-motion radiography

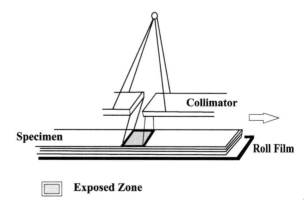

## 10.5  Auto Radiography

When the specimen being radiographed emits radiation of its own, it can be some-times used for taking a radiograph which is called auto radiograph. Researchers in biology use auto radiography for histopathology of sections of animal organs or tumors. In metallurgy, alloys made with isotope tracers can be tracked on auto radio-graphy of the polished section. Nuclear fuel elements are subjected auto radiography to detect inclusions of localized fissile particles which have higher radio-activity. However, this technique cannot be used for finding a mechanical defect in the compo-nent itself because the radiation emission is in all directions and a 'shadow' does not form.

## 10.6  Proton Radiography

Protons cannot be used like neutrons or X-rays because their attenuation does not follow the exponential law. Intensity drops abruptly at the end of its range in mate-rials. Thus the number of protons transmitted through a sample whose thickness is close to the range, is very sensitive to the exact thickness. Very good contrast sensi-tivity (sometimes < 0.1%) can be obtained. The other method is Scattered Proton radiography, which gives image with dark and light bands near the edges, like a edge-enhanced radiograph. Proton beams of high energy are not easy to find and hence the technique has little practical potential.

## 10.7  Radiation Gauging

Gauging is measurement of thickness, density or level. Gamma and Beta rays are used for this purpose. Both can be used in Transmission mode as well as Backscattered mode. IN specific cases neutrons are also used, particularly in oil wells. Basic layout is as in Fig. 10.5.

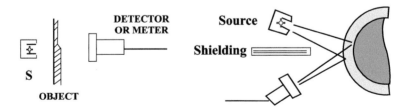

**Fig. 10.5**  Schemes for transmission and backscatter gauges

### 10.7.1  Beta Transmission

Sources: Sr 90, Y 90.

Applications: Thin sheet thickness—metallic or non-metallic (e.g. in sheet rolling or paper pressing).

### 10.7.2  Beta Backscatter

Backscatter intensity depends upon thickness of the scatterer layer and its atomic No.

Applications: Coating Thickness measurement, Paint thickness measurement. Backscatter gauge useful where the access is only from one side.

### 10.7.3  Gamma Transmission

Sources: Co-60, Cs 137 etc.

Applications:   Fluid Level detection.

Density or thickness of metallic plates or non-metallic sheets (e.g. rubber).

Density of contents in packages.

Maintenance needs like displacement of Trays in reaction towers of chemical plants and refineries.

### 10.7.4  Gamma Backscatter

For all single side access problems. Like pipe wall thickness, blockage or corrosion in pipeline, soil density, clad thickness etc.

## 10.8  Numericals

*Example 1*  A 30 mm thick part is radiographed at 700 mm SFD to find flaw depth by triangulation method. When source was moved by 250 mm, the flaw image moved by 3.5 mm. Find the flaw depth inside the part if specimen bottom to film distance is 5 mm.

*Solution*: The formula is distance $d = \frac{D\Delta F}{\Delta S + \Delta F}$ (Fig. 10.2, Eq. 10.1).

Substituting values, $d = \frac{700 \times 3.5}{250 + 3.5}$ i.e. 9.66 mm.

This is the total distance of flaw from the film. Considering 5 mm gap between the film and the specimen bottom, depth inside the job will be 4.66 mm from the bottom plane.

**Example 2** The above sample is radiographed with the double marker method but with film in contact with the job. Flaw image shift is 1.7 mm. Find the shift in top marker image.

*Solution*: Using Eq. (10.2), $h = t \times \frac{(\Delta f - \Delta BM)}{\Delta TM - \Delta BM}$, h being 4.66 mm and to find $\Delta TM$. Here $\Delta BM = 0$ as the film is touching the job.

$$\frac{h \times \Delta TM}{t} = \Delta F, \quad \text{or} \quad \Delta TM = t\Delta F/h$$

$$= 30 \times 1.7/4.66 = 10.9 \, \text{mm shift in Top marker image.}$$

## Model Questions

Q.1  A good contrast image is obtained using In-Motion radiography because of

   (a)  Speed of travel
   (b)  Use of collimators
   (c)  Filtration
   (d)  Motion unsharpness

Q.2  An important advantage of the Marker methods to find flaw depth when compared to triangulation method is that

   (a)  Source movement need not be measured in Marker methods
   (b)  Film is to be kept away from the specimen
   (c)  Calculations in triangular method are more complex
   (d)  Markers can be selected by the radiographer

Q.3  Which of the following is the most suitable application of the Beta transmission gauge?

   (a)  Metal coating thickness measurement
   (b)  Liquid level detection in chemical vessels
   (c)  Measuring and controlling paper thickness during paper manufacture
   (d)  Metal sheet thickness measurement during rolling

Answers are available in the section "Answers to the Model Questions".

# Chapter 11
# Filmless Options and Image Processing

## Contents

## 11.1   Filmless Modes

### 11.1.1   Why Real Time?

In spite of very good properties of X-ray films as the detection medium for radiography, screen fluoroscopy was brought in for the following benefits:

1. High speed of inspection in a continuous production environment (like automotive industry).
2. Instantaneous results as there is no processing required; from this fact comes the title "Real Time Radiography (RTR)" for this class of techniques.
3. Orientation of a defect with respect to radiation beam can be varied to bring it in to favourable location. This improves detection probability.
4. Positive Image—comfortable to view.
5. Ever rising costs of silver.
6. Real Time image is easily amenable to digitization and subsequent processing by a computer. (This consideration entered later, due to new developments in the field.)
7. No processing chemicals, eliminating environmental concerns.

© Ind. Society for Non-Destructive Testing 2024
P. R. Vaidya, *Guidebook for Radiography*,
https://doi.org/10.1007/978-981-99-8038-3_11

On the disadvantage side the factors are.

1. Resolution of film is difficult to match by most of the RTR options.
2. Very high cost of RTR equipments.
3. Operator has to have higher qualifications/experience as he directly interprets the image.

Of many developments that occurred over the time, the following have stayed.

A. Screen fluoroscopy
B. Image intensifiers

    B1.   Micro channel plates
    B2.   Video and CCD cameras to assist A and B.

C. Linear diode arrays
D. Flat Panel detectors (Digital Detector Arrays)
E. Photo-stimulable phosphor (or Image Plate).

All of them use **phosphor** materials directly or indirectly. Phosphors convert X-rays into visible light; either this light is directly viewed or converted to electrons to drive the other components. Roentgen had used barium platinocynide phosphor during the discovery of X-rays. The other popular phosphors are cesium iodide, zinc sulphide, zinc cadmium sulphide, gadolinium oxy sulphide, calcium tungstate, cadmium tungstate, cesium iodide etc. The following properties of the phosphors are critical for being effective.

- Spectral emission (wave length of light emitted)
- Stopping Power (which depends upon the atomic number Z and density)
- Conversion efficiency from X-ray to light (3000–5000 optical photons per X-ray photon)
- Persistence (how quickly the glow dies down, to be ready for the next event).

A. **Fluoroscopy**

ZnS, ZnCdS and $Gd_2O_2S$ are the currently used fluorescent screens. The colour of emission of phosphor should match with the peak of eye's response for better results. In dark room, the eye response shifts from yellow to green and hence green emission is preferred. The screen can be viewed directly with a lead glass barrier for radiation protection, or using periscopes, still cameras or video cameras (see Fig. 11.1).

Brightness in fluoroscopy is controlled by the thickness of phosphor layer and the grain size. Fine grain can have better resolution; unfortunately low thickness and small grain size yields poor brightness. Eye performs better at higher brightness. Thus resolution and brightness make opposing demands on fluoroscopy. Phosphors like $Gd_2O_2S$ bring a compromise to some extent but the limitation remains.

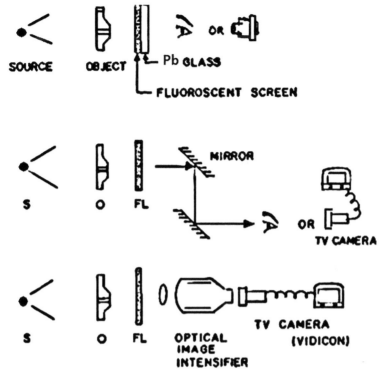

**Fig. 11.1** Ways to view fluoroscopic image

## B. **X-Ray Image Intensifier**

These are electronic tubes designed to increase brightness and overcome the above Paradox. The fluorescent layer is coupled with a photo-cathode which emits electrons on receiving the optical photon from the phosphor. In the modern versions, ZnCdS phosphor has been replaced by the scintillation crystals like Cesium iodide (CsI). This also emits optical light on receiving X-rays but the yield is much higher. Electrons from photo-cathode are focused and accelerated to give brighter picture on the phosphor at the exit end (Fig. 11.2). Image Intensifier also needs to be viewed by a camera.

### B1. *Micro-channel plate (MCP)*

This is an electron multiplier device. A glass tube with a coating of semi conductor layer is called a micro-channel. When an electron enters this channel which has a potential difference along the length, it multiplies in the numbers till it comes out on the other end (Fig. 11.3). By attaching it with a phosphor and photo-cathode even MCP can be used as an intensifier.

B2. *Cameras*

Video cameras are an important part of the RTR chain. Vidicon tubes of many varieties are available as per the need. Isocon and Orthicon are two more types of video cameras where the design is a variation from the Vidicon and are more sensitive. All of these are delicate in handling. Therefore the current trend is to use C-MOS and CCD (Charge Coupled Device) based cameras which are more rugged and light sensitivity is also better. There are special vidicons directly sensitive to X-rays which can be used for RT of small size objects.

## C. Linear Diode Arrays (LDA)

It is a type of digital detector array with a single line of detectors. It is also called **Line Detector**. A silicon based device gives out electron pulse when coupled with a phosphor. An array of small size detectors make a line scanner. The scanned data is kept in Frame-store till the full image is collated from individual rasters. Here the scattered radiation is negligible because of slit scanning arrangement. Mainly used for baggage inspection at airports (Fig. 11.4).

## D. Flat Panel Detectors (Digital Detector Arrays)

A flat panel detector is a 2-dimensional array of tiny X-ray detectors. There are 2 types of DDA's available commercially, called direct conversion and indirect conversion detectors. Indirect conversion detector consists of phosphor layer such as Cesium iodide or Gadolinium Oxysulphide which converts the X-ray or gamma radiation to visible light by scintillation process. The light is converted to electrical signal by using Amorphous Silicon diode array. Charge Coupled Devices (CCD) and Complimentary Metal Oxide Semiconductor (CMOS) are other forms of readout devices used for converting light to electrical signal.

**Fig. 11.2** X-ray image intensifier

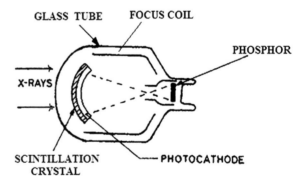

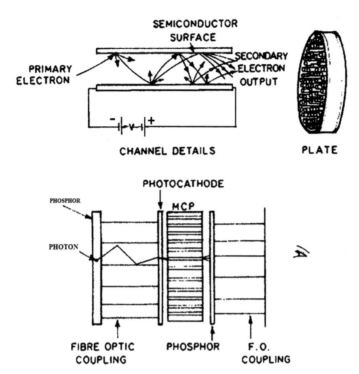

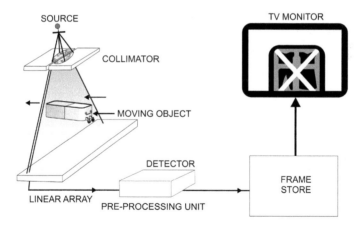

**Fig. 11.3**  A typical micro channel, MCP and its use as image intensifier

**Fig. 11.4**  Linear diode array

In direct conversion detectors, the X-ray or gamma photon is directly converted to electrical signal by principle of photo conduction. Amorphous Selenium array is an example of Direct conversion detector. The converted discrete array of analogue signals from DDAs are subsequently digitized and transferred to computer for display of image. The technique is also digital radiography (**DR**).

E. **Image Plates (Photostimulable Phosphors)**

Certain phosphors like BaFBr (Eu) have a property that they store the electrons emitted during the interaction with ionizing radiation in their colour centres (called F-centres) in the lattice. When they are stimulated using laser light, they give out optical light in the violet range to come to ground state. This light is proportional to the radiation absorbed and thus can create a digital map of the image. Image Plates are available in rigid and flexible forms. Respectively the flat bed and drum type scanners are used to scan the plates after the exposure. The image is erased using bright light; the plates can be used for large number of times repeatedly.

Just like films the plates need (electronic) processing and hence these are not 'real time' detectors. Image Plate systems are also called 'Storage Phosphors' or "Computed Radiography (**CR**)", though there is no sound logic behind the latter name as the computers are used for other detectors also.

## 11.2   Computed Tomography (CT)

A conventional radiography gives a plan view of the object being radiographed. The depth information is missing, which should be gathered by radiography in transverse direction. But there are certain objects which can not be X-rayed in the transverse (axial) direction to obtain a plan view or cross sectional view. Computer assisted tomography can help in such situations. Here the object is radiographed in transverse direction and digital data is collected on the detectors placed diametrically opposite. Large number of exposures are taken by rotating the object through small angles. The entire data set so collected is then used for re-construction of image using mathematical algorithms. These algorithms have the capacity to create a cross-sectional image from the data collected in radial direction. In fact using the computer system, any image in a desired plane can be retrieved from this data. CT has applications in aerospace and other engineering industries (Fig. 11.5).

## 11.3   Image Processing and Analysis

Image processing is to modify the image so that it can give the detailed information. Image Analysis is to extract that information from the image.

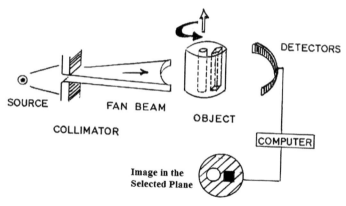

**Fig. 11.5** Explanation of CT process in the fan beam mode

This image can be on film or on filmless medium like fluorescent screen, video monitor or digital data on hard disk. For all cases there are 4 main attributes:

- Density (or brightness)
- Contrast
- Resolution
- Signal to noise ratio.

If the image is not capable of giving the required information, due to deficiency in one or more of these factors, it can be processed for the improvement.

In the modern age, processing and analysis are by digital means done through a computer. (There are optical means for processing, now not used.) Therefore analog images from the film, fluorescent screen or Image Intensifier output, need to be digitized first. If all of them are viewed by a video camera, the output can be easily digitized. For films there is one more way to create digital image i.e. by scanning them on rotating drum or on a flat bed scanner and collecting the transmitted light point by point.

In the digital plane the smallest element of image which is at a unique gray level is called **Pixel** (i.e. picture element). The image attributes are measured by different units.

Brightness: By gray scale numbers. The range for dark to bright can be divided in 64. 256, 512 etc. segments. In 0–156 scale 0 is the darkest point, 256 is the brightest point.

Contrast: Difference in brightness.

Resolution (Sharpness): In terms of spatial frequency i.e. line pairs/mm.

As contrast and resolution are interdependent, a new parameter called Contrast Transfer function or Modulation Transfer Function (**MTF**) is defined for the whole range of frequencies. (It is actually the property of imaging system and not the

**Fig. 11.6** A typical MTF
curve

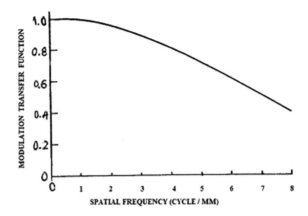

image. But image is anyway the outcome.) At higher frequencies, the contrast is not reproduced faithfully. Thus MTF shows up in a typical curve like in Fig. 11.6.

### 11.3.1   Processing Filters

As far as computer is concerned, an image is nothing but a collection of grey level data with the coordinates of the respective points. Hence the manipulation of an image will need to alter the grey level value of some or all the pixels. The characteristics of the image and the shortcoming will decide the nature of modification required. It is achieved by the algorithm used for this purpose. If the cause of shortfall is known, the processing is like reversing the cause and so called '**Restoration**'. Or else it is "**Enhancement**". The algorithm in mathematical terms is the 'filter'. It operates on the image either point by point (pixel by pixel) or to the image as a whole. The former method is called space based technique (examples—average, median filter, median etc.) The other one is called frequency domain technique (e.g. histogram modifications). More often a combination of filters is needed to achieve the result.

### 11.3.2   Analysis

Many times the analysis may need a preliminary processing of the image to make it suitable for analysis. For example, shade correction or segmentation may be needed before one counts porosities or measure their total area. In assemblies with fixed geometries, comparison of two images can find missing components or a foreign object.

**Nugget**

The humble radiography film actually performs three functions at a time: it **detects** the radiation, **records** it and then **displays** the image. This can be appreciated when one thinks of real time or digital radiography (like Image Plate). There the detector is different, the image is recorded in computers and it is displayed by the video monitor. All these functions are done by the single film in a thickness of just 100 μm!!

**Model Questions**

Q.1  Screen fluoroscopy is one of the methods in Real Time RT family. It suffers from one disadvantage

(a) High brightness
(b) Job can be positioned as desired
(c) Immediate results
(d) Poor resolution capability

Q.2  Which of the following properties does not apply to Flat Panel detectors?

(a) Uses amorphous silicon based diode array
(b) Have some dead pixels
(c) Image erased by a bright light for reuse of the detector
(d) Uses phosphors like Cesium Iodide or Gadolinium Oxy Sulphide

Q.3  The only software based technique which can distinguish features lying one behind the other in the direction parallel to the radiation beam

(a) Linear diode array
(b) Computed tomography
(c) Image phosphor plate
(d) It is not possible to do so

Answers are available in the section "Answers to the Model Questions".

# Chapter 12
# RT in Different Industrial Sectors

## Contents

Basic radiographic technique remains same when it is applied in any industrial sector like Chemical, Space, Marine etc. However there are certain specific variations which are typical to that particular industry. In the following sections, examples of such practices are given. Apart from these each industry will have the routine RT applications anyway.

## 12.1  Aerospace and Aeronautics

An Application typical to the space technology is radiography of rocket motors. They come in various sizes. While small ones are tested with conventional X-ray units and with conventional techniques, the larger ones (> 1 m dia) need Linear Accelerators or High activity isotope source. Rocket motor is basically a thick walled metallic cylinder filled with propellant material, called the 'grain'. First the cylinder is tested for its basic mechanical integrity. As the propellant material is highly explosive, the assembly needs to be handled carefully after the propellant is filled (Fig. 12.1).

There is an insulation lining of a resin like material between the metallic shell and the propellant. Tangential RT is used to check the quality of bond on either side of the insulation. The other test is to check the casting quality of the propellant grain, to ensure that it is free of cracks, voids, inclusions or deformation in the central

© Ind. Society for Non-Destructive Testing 2024
P. R. Vaidya, *Guidebook for Radiography*,
https://doi.org/10.1007/978-981-99-8038-3_12

**Fig. 12.1**  A cross section of
the rocket motor

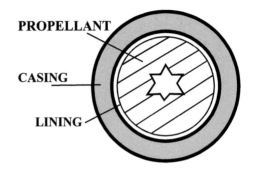

port. The configuration of light material encased into a heavy metallic container is a typical candidate for neutron radiography. To obtain an intense source of neutrons often the LINAC is modified to become neutron source. A beryllium disc in the path of X-rays gives copious supply of neutron by ($\gamma$, n) reaction for neutron RT.

Beryllium (Be) is also used in the space industry for making mirrors and gyroscopes. X ray radiography of vacuum hot pressed Be blocks is done taking all the precautions needed for the low energy radiography.

### 12.1.1  Aeronautics

Aircrafts are tested before a flight or as a part of periodic inspection. Most of the components here are made of aluminum/titanium or their alloys. RT is done for Engine, Frame structures, Landing gear related parts, Fuselage, Wings and Rudders etc. The last two are made of Honeycomb structures, with covering panels attached on both sides using brazing or bonding. As per the part being tested, energies from 100 kV X-rays to Ir 192 are employed in RT.

Typical defects in Honeycomb structures are delamination, inclusions, fibre breaks, water ingress, debris, corrosion, compressed panel, damaged panel, issues related to composite layer bonding like voids/pores, resin rich or starved areas etc. Radio-opaque liquids (like zinc iodide or carbon tetra chloride) can be used to highlight defects behind thin sections.

## 12.2  Oil and Refineries

As compared to radiography, use of gamma gauging and radiometry techniques is more prominent in this industry. During the oil well logging, the gamma and neutron back scatter gauges are sent down in bore holes to monitor the strata surrounding the hole. This can distinguish between water, oil or sea water. Similarly the nature of rocks can also be judged.

**Fig. 12.2** Depth of corrosion

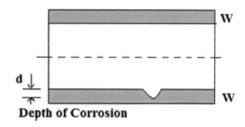

**Depth of Corrosion**

In the refineries and chemical plants gamma scanning is used to monitor the processing columns and fractionating towers. The technique is also used for troubleshooting, in finding liquid/vapour boundary or when process trays move away or shift from their locations.

In the case of pipelines, if radiometry detects corrosion or blockage, radiography is useful in identifying the exact location and nature of corrosion/obstruction. For cross country lines with long stretches, travelling Crawlers are used with either the X-rays or isotope as the source.

## 12.2.1 Assessment of Corrosion Depth

For measurement of residual wall thickness after corrosion in a pipeline, Ultrasonics is the suitable method. However if the pipe is hot or there is an insulation, one uses RT for the purpose. It can be used in two ways; taking tangential image or in 'double wall—double image' (DWDI) mode.

### 12.2.1.1 Tangential RT (TRT)

As discussed in Chap. 7, TRT provides a longitudinal profile of the pipe wall, from which depth of corroded pit can be assessed (Fig. 12.2). Correction for the magnification due to distance of the film from the pipe centre may be applied. To apply this method one needs to first locate the corrosion area and bring it to the tangent location.

### 12.2.1.2 DWDI Mode

In normal DWDI radiography the corroded patch will show higher density compared to the neighbouring area. Using H&D Curve one can make a rough estimate of the depth. Assume that the densities in surrounding area and on the patch are respectively $D_1$ and $D_2$, $(D_2 > D_1)$. The Relative Exposures on the film characteristic curve are $E_1$ and $E_2$. The ratio

$$\frac{E1}{E2} = \frac{2^{\frac{x1}{HVL}}}{2^{\frac{x2}{HVL}}} = 2^{\frac{\Delta x}{HVL}} \tag{12.1}$$

where $\Delta x$ is $(x_2 - x_1)$.

As $E_1/E_2$ as well as HVL for given energy will be known, $\Delta x$ can be found out. That is same as corrosion depth 'd' in the figure. This method is very approximate because RT is usually done in a broad beam geometry (and not narrow beam geometry); thus the exposure values are not faithful to thickness values and the answer in Eq. (12.1) can be only an estimate. It may be just good enough for the engineering decisions on the shop floor.

## 12.3   Nuclear

Very wide variety of specimen types occur in this field. Apart from conventional vessels and pipelines being tested for weld and casting integrity, the weld integrity in nuclear reactor **fuels** is an important area. Usually these welds are of very fine dimensions and need special techniques to carry out meaningful radiography. In majority of cases tangential radiography technique is used for the purpose. Use of half or full compensating blocks becomes necessary in view of very thin wall of the tubes, to avoid burnout (Fig. 12.3). Tangential RT provides an image like a vertical cross section of the weld and shows up important flaw like lack of fusion or root burning with accuracy in position.

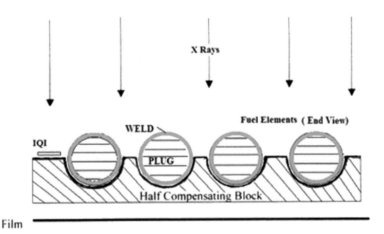

**Fig. 12.3**  Example of half circular compensating block

## 12.3.1 Radiography of Radioactive Object

During the In-service-Inspection or maintenance in a nuclear reactor, a need arises to radiograph objects having residual radioactivity. The primary difficulty in doing this is that the radiation can cause blackening of the film directly without relation to the flaws, if any. It thus gives fog and also reduces contrast. To overcome this, the transfer technique of neutron radiography (NR) is developed and is used mainly for irradiated fuel elements. However NR is not always a quick and ready method available. And it is indeed possible to take X- or gamma radiograph of the radioactive object with moderate activity.

Following steps are taken to reduce/eliminate the effect of radiation from the object on the film. The need is to have a short exposure time so that the minimum blackening happens. For this,

(i) Use X-rays rather than gamma sources which usually have longer exposure times.
(ii) Use a slow film so that the density obtained by 'fog' radiation remains low. Useful radiation from the source is much higher and can contribute to major part of density obtained (blackness).
(iii) Use highest compatible kV (within sensitivity limitations) to reduce exposure time further.
(iv) Keep a small gap between the object and the film using soft material like thermocol etc. to reduce the background radiation. Use direct enlargement, is possible, using mini or microfocus units.
(v) Keep SFD low for the same purpose but within acceptable $U_g$ limits.
(vi) Allow minimum time of contact between the film and the object by quick operations.
(vii) Usual radiation safety of the operating staff is to be taken care with respect to specimen's radioactivity.

## 12.4 Marine and Off-Shore

This applies to Ships, submarines and offshore structures. RT used in two stages. (i) During manufacture (ii) During service: In the dry dock, At sea.

(i) Components tested are Hull (Usually welded by SAW technique), Propulsion Unit, Piping, Castings/Forgings (for Anchors, Rudder Bearings, Propellers, Shafts, Brackets, Gun Mountings etc.)
(ii) During service, MPT and UT are more frequently used tests. Though RT is not a popular choice, it is used for

   (a) checking corrosion or wall reduction in pipes $< 2''$ dia where wall thickness testing by UT is unwieldy; this is done by tangential RT

(b) checking the integrity of Riser pipes which run from the sea bed till the offshore platform and carry the product or water for injection. They can be rigid or flexible; the flexible ones are made of layers of Composite material and steel wires alternatively. Even if RT is not a very satisfactory technique but is used in absence of a better option.

**Model Questions**

Q.1  Honeycomb structures are typical to this industry

   (a)  Space
   (b)  Aeronautics
   (c)  Chemical and refinery
   (d)  Marine and offshore

Q.2  Which action will not help the cause of good radiography of a radioactive object?

   (a)  Use quick operations to minimize the time of contact between the object and the film
   (b)  Use neutron radiography with transfer technique
   (c)  Keep SFD large to improve geometrical unsharpness
   (d)  Use X-rays instead of gamma rays to have shorter exposure times

Answers are available in the section "Answers to the Model Questions".

# Chapter 13
# Radiation Safety

## Contents

Basic units for radiation measurement were mentioned in Chap. 1. Then the Equivalent Dose was defined as the Absorbed Dose from a type of radiation $D_R$ x weighting factor of that radiation $W_R$.

$$\mathbf{DoseEquivalent} \ = \ \sum \mathbf{D_R W_R} \tag{13.1}$$

$W_R$ varies as per the linear Energy Transfer (LET) of each radiation in the material or the body tissue (Table 13.1).

In the same manner, different tissues are sensitive to radiation in different measure. There are Tissue weighting factor ($W_T$) to grade them. Effective dose for whole body will be summation of such local organ doses.

$$\mathbf{EffectiveDose} \ = \ \sum \mathbf{H_T W_T} \ \text{(Sieverts)} \tag{13.2}$$

where $\mathbf{H_T}$ is the dose equivalent, $\mathbf{W_T}$ is tissue weighting factor (Table 13.2).

© Ind. Society for Non-Destructive Testing 2024
P. R. Vaidya, *Guidebook for Radiography*,
https://doi.org/10.1007/978-981-99-8038-3_13

| Type of radiation | Weighting factor $W_R$ |
|---|---|
| X or gamma rays I all energies | 1 |
| Electrons | 1 |
| Protons > 2 meV | 5 |
| Heavy nuclei including Alpha particles | 20 |
| Thermal neutrons | 5 |
| Fast neutrons (>20 meV) | 20 |

**Table 13.1** Weighting factors for different radiations

| Oragn | $W_T$ | Organ | $W_T$ |
|---|---|---|---|
| Gonads | 0.2 | Liver | 0.05 |
| Bone marrow | 0.12 | Breast | 0.05 |
| Lungs | 0.12 | Thyroid | 0.05 |

**Table 13.2** Tissue weighting factors

## 13.1   Biological Effect of Radiation

Radiation produces free radicals in the cells and creates damage. The exposure can be two types, acute (large dose at a time) and chronic( distributed over long time). The effects are also of two types, stochastic and non-stochastic. Non-stochastic effects are seen only on the person receiving the dose. The stochastic effect is statistical in nature and can be seen also on the future generations of the affected person. Effects which have potential to be seen in future generations are called **Genetic Effects**. There are no thresholds for such effects. Most prominent chronic dose comes from the natural background radiation which is **2–3 mSv per year** (Table 13.3).

## 13.2   Measurement

Radiation measurement is required for personnel safety as well as for area monitoring. Film badge, pocket dosimeter and TLD badge are personnel dosimeters. They measure **total dose** (mR/mSv), whereas area monitor measure **dose rate** (mR/hr or mSv/hr).

There are 4 classes of measuring devices viz.

Gas filled counters, Scintillation detectors, Solid state detectors, Photographic film.

**Table 13.3** Effects of acute dose

*(a) whole body irradiation*

| Dose range | Immediate effect |
|---|---|
| < 100 mSv (< 10 rem) | None detectable |
| 0.1–0.25 Sv (10–25 rem) | Chromosome aberrations, recoverable |
| 1–3 Sv (100–300 rem) | Nausea, vomitting, diarrhea (NVD) loss of appetite. Recovery probable |
| 3–5 Sv (300–500 rem) | NVD plus fever. Radiation sickness. Recovery probable |
| > 5 Sv (> 500 rem) | Death within a few days to weeks |
| 4–6 Sv | LD50/30: Chance that 50% of exposed persons will die in 30 days |

*(b) Local dose on organs*

| Dose | | Organ | Effects |
|---|---|---|---|
| Sv | rem | | |
| 0.15 | 15 | Gonads—men | Temporary sterility |
| 0.30 | 30 | Blood | WBC count drops |
| 0.5 | 50 | Blood | RBC count drops |
| 3 | 300 | Skin | Temporary epilation |
| 6 | 600 | Skin | Permanent epilation |
| 1.6 | 160 | Gonads—women | Temporary sterility |
| 2.5–6 | 250–600 | Gonads—women | Permanent sterility |
| 4 | 400 | Gonads—men | Permanent sterility |
| 5 | 500 | Eye lens | Cataract after 5–10 years |
| 6 | 600 | Skin | Skin erythema |
| 10 | 1000 | Skin | Burns, wound, tissue death |

## 13.2.1 Gas Filled Counters

The devices consist of an air chamber and two electrodes as shown in the Fig. 13.1. A potential is applied between cathode wire and anode body. On exposure to radiation, ion pairs are produced in the gas, which drift to positive and negative electrodes, respectively, as per their charge. The resultant pulse is extracted across an impedance and counted or measured on a meter. Depending upon the voltage applied the behavior of ion pulse varies and so does the property of the detector. See Fig. 13.2 for the regions of applied voltage in which they operate. Following are gas filled detectors with increasing order of applied voltage at which they work.

(i) Ionization Chamber

Very versatile device, with very small to a large size possible. Mainly air filled though different gases are suitable media. No gas multiplication takes place in ionization chamber. It is preferred for measuring high level of radiation. These devices have a

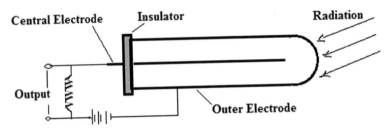

**Fig. 13.1** A typical gas filled detector

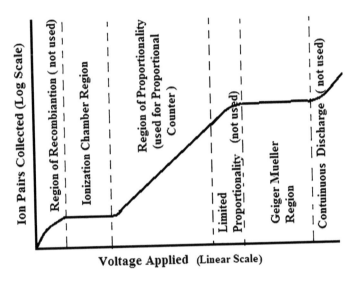

**Fig. 13.2** Regions of operation for gaseous detectors

good uniform response to radiation over a wide range of energies. Pocket dosimeter is an ionization chamber.

(ii) Proportional Counter

It is an ion chamber generally filled with P-10 Gas (90% Argon and 10% Methane) operated at mid voltage range. Primary ion pair produced makes further ionization along its path. This is called as gas multiplication process. Pulse height is proportional to the energy of radiation trapped. This helps determining energy of radiation along with the intensity. BF3 counters for neutron detection are the proportional counters by design.

(iii) Geiger Mueller (GM) Counter

Widely used device, though response is slower than proportional counters. Rugged in operation. The entire volume of gas inside get ionized due to gas multiplication process. It is widely used in Radiation survey and Area zone monitors. The limitation is that it cannot differentiate energy and it cannot be used for higher radiation field because of dead time.

If very thin window is provided, can be used for counting beta particles also.

Area Survey meters and stationery Area Zone monitors usually use GM counters as detectors.

## 13.2.2  Scintillation Detectors

These devices use certain single crystals as the detector in which a light flash occurs on absorption of beta or gamma particles. NaI and $C_SI$ crystals are used for gamma whereas Anthracene for beta rays. The light flash is collected by the photo cathode on a photomultiplier tube (PM tube). Electrons emitted by photocathode are multiplied by the PM tube to result in a pulse at the anode.

These are very sensitive and fast detectors which can discriminate energy of radiation. Occasionally liquids can also be used as Scintillation detectors.

## 13.2.3  Solid State Detectors

(i)  Thermoluminiscent dosimeters (TLD)

Some compounds like $CaSO_4$ (Dy), LiF etc have properties that can trap the electrons released due to radiation absorption. These electrons are later released if the substance is heated. Their amount is proportional to the intensity of the absorbed radiation. TLD is a popular device for personnel dosimetry for radiation workers.

(ii)  Semiconductor based detectors

Junction type detectors of Germanium and Silicon (Li drifted, both) are very popular because of very high sensitivity and energy discrimination. They are delicate and complex in operation, hence not used in field service.

## 13.2.4  Film Badge

Photographic films can be used for detection as well as measurement of radiation. As the optical density is proportional to the radiation quantity, Film Badge were used as personnel dosimeters earlier and have been now replaced by TLD Badges.

## 13.3  Dose Limit Regulations

ICRP has recommended following dose limits.

### 13.3.1  For Humans

|  | Occupational worker | Members of public |
|---|---|---|
| Whole body | 20 mSv (2 rem)/year, averaged over 5 years. 30 mSv in a single year | 1 mSv in a year. (Extended to 5 to 50 mSv in special circumstances) |
| Eye lens | 150 mSv (15 rem) | 15 mSv (1.5 rem) |
| Skin | 500 mSv (50 rem) | 50 mSv (5 rem) |
| Hands/ foot | 500 mSv (50 rem) | 50 mSv (5 rem) |

However the principle of ALARA (As Low As Reasonably Achievable) has to prevail always.

### 13.3.2  For Cameras

|  | On the surface | At one metre distance |
|---|---|---|
| Portable | 2 mSv (200 mR)/hr | 0.02 mSv (2 mR)/hr |
| Mobile | 2 mSv (200 mR)/hr | 0.05 mSv (5 mR)hr |
| Fixed | 2 mSv (200 mR)/hr | 0.1 mSv (10mR) /hr |

## 13.4  Transport Index (Ti)

Transport Index is the number expressing maximum radiation level in mrem/hr at 1 m from the external surface of the package being transported. Its permissible range is 1 to 10. For portable, mobile and fixed cameras, TI is respectively 2, 5 and 10.

The category of radiation label is to be selected based on TI. For radiography camera in transit label of Category III—Yellow is required.

**Table 13.4** Half value layer for different materials (in mm)

| Radiation | Effective energy (keV) | Pb | Concrete | Fe | Uranium |
|---|---|---|---|---|---|
| 200 kV X rays[a] | 130 | 0.5 | 25 | – | – |
| 400 kV X rays[a] | 240 | 2.4 | 36 | – | – |
| 1 MV X rays | 660 | 7.5 | 50 | | |
| Ir $^{192}$ | 440 | 5 | 45 | 12.5 | 3.1 |
| Cs $^{137}$ | 660 | 8 | 50 | 16 | – |
| Co $^{60}$ | 1.25 | 12 | 63 | 21 | 7 |

[a] HVL values for x rays can vary greatly depending upon the inherent filtration
(Tenth Value layer thickness can be found from **TVL = 3.322 HVL**)

## 13.5  Shielding and Cordoning

In order to keep dose within limit for the radiation worker as well as any one from the public, cordoning is required for the area where radiography is being done. If adequate distances are not available one may need additional shielding. While cordoning it will be always considered that the members of public will be at the cordon fence and not the radiation worker).

Hence the dose limit to be achieved will be 1 mSv in a year or 1/50 mSv (= 0.02 mSv) in a week. Cordon Distance now can be found from the Source Activity, RHM of that Source and weekly work load data (No. of exposures x duration of each exposure).

Use the data of Half Value layer HVL (or half value Thickness HVT) given in Table 13.4 to solve numericals given below.

## 13.6  Numericals

***Example 1*** You are working with 200 kV x-rays and have concrete blocks as the shielding material.

(a)  What thickness of concrete block will be needed to reduce radiation intensity by 100 times?
(b)  By 40 times?

*Solution*:

(a)  At 200 kV concrete HVT is 25 mm. Its TVT should be 3.322 × 25 = 83 mm. To reduce intensity by 100 times, one needs 10 TVT of material. Hence **830 mm** of Concrete block will be needed.
(b)  40 times reduction can be achieved by 2 HVT and one TVT (4 × 10 = 40). Therefore (25 × 2) + 83 mm of concrete should be needed. This is **133 mm**.

*Example 2* What is the cordon off distance when Ir 192 source of 555 GBq (15 Ci) is used in a busy workshop? Workload is 10 h in a week.

*Solution:* A member of public can take a dos of 2 mrem/week (from 100 mrem/year).

In the present case it will be given over 10 h. Hence the dose rate at the cordon fence should be less than or equal to 2/10 i.e. 0.2 mrem/hr.

As Ir RHM is 0.5,

15 Ci Ir Source Will Give $15 \times 0.5 = 7.5$ R/hr at 1 m. I.E. Dose of 7500 mrem/hr

$\frac{Dose\ 1}{Dose2} = \frac{(distance\ 1)^2}{(distance\ 2)^2}$ here distance 1 is 1 m.

Hence $(distance\ 2)^2 = $ dose 1/dose 2 $= 7500/0.2 = 37,500.$

Or distance 2 $= \sqrt{37,500} = 194$ m is the cordon distance.

*Example 3* In the Example 1, the distance available at site is only 97 m. Therefore radiographer wants to put a Lead sheet curtain around the source. What thickness of Pb will be required?

*Solution:* Only 97 m are available instead of 194 m. This is half the distance. Hence the dose level at cordon will be $(2^2 = )$ 4 times higher than achieved earlier. A Pb sheet of 2 HVT will bring down the radiation level to ¼ again. HVT for Pb for Ir 192 is 5.5 mm. Hence 11 mm sheet of lead will be required.

*Example 4* An enclosed radiography installation is required to use Co-60 source with maximum activity of 185 GBq (5 Ci). The 6 m × 6 m room has a concrete wall 40 cm thick. Persons working outside are radiation workers. Is this wall adequate if the source is used only in the centre of the room and for 5 h a week? (Concrete TVT is 190 mm, HVT 63 mm).

*Solution:* Source gives out $1.3 \times 5 = 6.5$ R/hr at 1 m. Source is at 3 m distance from the wall. Hence dose rate at the wall (inside the room) will be $6.5/3^2 = 6.5/9 = 0.73$ rem/hr.

The wall is 400 mm i.e. $400/63 = 6.35$ HVTs, therefore its reduction factor is $2^{6.35}. = 81.6$

Dose rate will reduce by this factor. i.e. 730/81.6 mrem/hr $= 8.95$ mrem/hr. Thus weekly cumulative dose will be $8.95 \times 5 = 44.7$ mrem. Radiation workers are permitted 40 mrem in a week. Thus the installation is NOT adequate.

*Example 5* An unknown source gives a dose of 100 mr/hr at 5 m from the exposure site. By inserting a steel plate of 25 mm, the dose reduces to 25 mr/hr. What is the source and what is its activity?

*Solution:* Inserting a steel plate reduces the dose by a factor of 4. So the 25 mm plate is 2 HVTs or 1 HVT is 12.5 mm. From Table 12.4 in Chap. 12 this has to be Ir 192 source.

Inthat case it 1 Ci should give 500 mr/hr at 1 m. or $500/5^2 = 20$ mr/hr at 5 m distance.

Here the dose rate at 5 m is 100 mr/hr. Thus the source is 5 Ci, Ir 192.

**Example 6** If a concrete room is approved for Co-60 operations of a particular activity. What level of activity of Ir 192 source can be handled in that room?

*Solution*: HVT of concrete for Co-60 is 63 mm.
HVT of concrete for Ir 192 is 14.5 mm.
The ratio 63/14.5 = 4.3.
Hence 4.3 times the activity of Co -60 permitted in this room can be handled in here for Ir 192 source.

**Example 7** A cylindrical SS vessel (4 m dia, 50 mm thick) is radiographed using 178 GBq (4 Ci) Ir 192 source in panoramic mode. What will be the dose rate at the cordon 2 m away from the vessel?

*Solution*: Side view of the operation will be as given below.

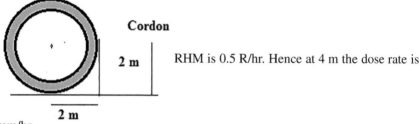

RHM is 0.5 R/hr. Hence at 4 m the dose rate is 2 rem/hr.

But the cordon is at 4 m from the source. So the dose rate in absence of the vessel will be $2000/4^2$.

The vessel is 50 mm or 4 HVT of Steel. It will reduce the dose by $2^4 = 16$ times further. i.e. it will be $\frac{2000}{16 \times 16} = 7.8$ mr/hr **Ans**.

**Example 8** Radiation level due to Co-60 source 30 mr/hr at 3 m after going through 1 TVT + 2 HVT of material. Find the activity of the source.

*Solution*: RHM is 1.3 for Cobalt. At 3 m it should give 1300/9 mr/hr for every Curie.
Additionally 1 TVY + 2 HVT reduces dose by $(10 \times 4 =) 40$ times.
i.e. $\frac{1300}{9 \times 40} = 3.6$ mr/hr every Ci. Dose rate here is 30 mr/hr.
Hence activity is 30/3.6 = 8.3 Ci.

## Model Questions

Q.1 which organ is considered sensitive to radiation damage and is given the highest tissue weighting factor?

    (a) Lungs
    (b) Gonads
    (c) Liver
    (d) Skin

Q.2 Geiger Muller counter has all but one of the following characteristic. Identify that one.

(a)  Rugged in operation
(b)  Has the response proportional to radiation energy
(c)  Widely used in Area Survey Meters
(d)  Can detect both, beta and gamma particles

Q.3  A carpenter is working in an area near to radiography enclosure. He worked for a week and received a dose of 80 mR (0.8 mSv). He has no chance to receive any more radiation in the next year.

(a)  He should be sent for a blood examination for chromosomal damage
(b)  As his dose is below the permissible limit, nothing need be done
(c)  He has crossed the stipulated limit, so the radiographer should beheld responsible
(d)  He should be informed of his dose and that it is within limits. But he should not take new dose more than 0.2 mSv during the next year.

# Chapter 14
# Codes, Standards and Procedure Preparation

## Contents

When radiography is done as a part of a broader Quality Assurance Plan, it is necessary that it is done as per some predetermined plan and is repeatable. To meet this requirement, a procedure is laid down for RT for the specific component (or system) with its acceptance specifications in mind. This document itself is called **'the procedure'**. It can be used in-house by an organization for its own QC needs or can be a bilateral document agreed upon by the parties if RT is being done by a radiography agency for the manufacturer of the components.

The procedure is based on the requirements of testing as mentioned in one of the three documents viz. the specification, the Standard or a Code section. It depends upon which of the three governs the contract between the manufacturer and purchasing agency. This way, the Procedure is a document connecting the Specification (or a Standard) followed during the fabrication and testing with the shop floor practices. It is prepared by the level 3 person responsible for the testing, approved by his organization and agreed upon by all concerned agencies. Specific part of the Level 3 examination needs candidates to write a procedure for a given component with the given specification/standard.

© Ind. Society for Non-Destructive Testing 2024
P. R. Vaidya, *Guidebook for Radiography*,
https://doi.org/10.1007/978-981-99-8038-3_14

## 14.1   Specification

For any component or an equipment, a detailed technical characteristics are listed in the specification. It can be dimensions, mechanical properties, metallurgical properties or physical properties like viscosity, density etc. These are usually measureable entities and decided by the designer for the satisfactory performance of the component. The procedure can select the correct technique to meet the needs of the specifications.

## 14.2   Standard

This is a document which lays down practices or technical requirements for a product, process or a service. Unlike Specifications, this can recommend non-measurable like Safety, Cleanliness, good practice etc. When related to materials, it can recommend a method of testing. Standards can be mandatory or in a nature of a Guide.

## 14.3   Code

A Code is also a type of a Standard or a collection of Standards. It has an authority like a Law. Usually Codes deal with the equipment or component in a comprehensive way; it deals with its design, material selection, fabrication and testing. Very famous example of a Code is ASME's Boiler and Pressure Vessel Code, though there are French and German Codes also. ASME adopts ASTM Standards when required, by renaming them. (For example E-142 becomes SE-142 etc.). *If no particular Specification or Standard is referred to, then the Procedure can be written to adhere to a Code in general.*

A typical Standard is given in the Annexure 1 as a basis for the Procedure Writing. This can also be used to answer the questions at the end of this chapter. Various Codes like ISO 9712, IS 18305 or BS 3998 have given guidelines as to what points should be included in the Procedure as well as in the Instruction. Some administrative points are also included (like a mention of connected documents or safety requirements etc.).

## 14.4 Preparation of the Procedure

A procedure can be very elaborate or a short one. But it has to contain those factors which can *influence the Quality of radiograph and the Quality of results.* Following points are essential for a RT procedure; though they can come in any order or grouping or sub-headings.

1. Scope

    1a. description of specimen, material etc.

2. Applicable Documents/standards
3. Qualification of personnel
4. Equipments used
5. Technique Details or Test parameters

This section carries large component of marks.
   One may refer to Instruction document here or fully describe the following:

   (i) Shooting geometry (Double Wall/Single Wall, Direction and SFD)
   (ii) Exposure parameters like kV or source, time)
   (iii) Film/screen used
   (iv) Geometric unsharpness
   (v) IQI
   (vi) Processing
   (vii) Density range

6. Acceptance standard and disposal of unaccepted parts. There can be discussion on repair and retesting of repaired parts etc.
7. Documentation—Record, Report, Storage of films etc.

   The list is not exhaustive and more points can be added as per the nature of the job or contract terms.

## 14.5 Technique Details

Before a procedure is written down, one has to make decision about technique employed, source, shooting direction etc. In real life this technique is decided by taking trial exposures when needed. For the purpose of Certifying Examination, certain things (like no. of shots) can be assumed. However Ug values must be checked to meet specification values. Often the question paper lays conditions; then the technique should satisfy those conditions. For example, if the rider in the question says "…….. with best possible sensitivity" then one should opt for slower films and/or select highest practicable SFD or lowest possible kV etc. Instructions in Chap. 7 can be revisited here.

**Different variations** are possible in the style of writing the procedure. For example, Equipment details can be a part of 'Technique' clause. If the purchaser's specification calls for a particular surface preparation (e.g. weld bead flushed) or stipulates maximum KV etc., such points can be included in the procedure for special emphasis. Very often a sample of Report sheet is attached with the procedure. It is a good practice but may not be essential for the exam.

Similarly if RT is to be done at a particular stage only (say, only after post weld heat treatment) then a clause on "Time of examination' can be included before Clause 4. For very complex and critical assemblies, sequence of welding and radiography may be important. For further clarity this can be mentioned as it often influences the choice of technique employed.

Annexure 2 and 3 here gives two sample procedures, one each for Welds and Casting as a general guideline. The weld procedure is based on the mock Standard given in Annexure 1 titled 'Sample Std 1: 2023'. Procedure for Casting in Annexure 3 is based on ASME Code. The technique details as given in Clause 5 (Annexure 2) and Clause 6 (Annexure 3) are the heart of the procedure and carry a large portion of marks. They are decided based on the demands given in the Sample Std 1. To provide a good sample for practice to the students, Annexure B has included all the likely variations in pipe radiography techniques.

## 14.6   Sample Questions

The specific examination also includes some questions on the interpretation of the standard supplied. Following questions are based on the Sample Std 1 given in Annexure 1.

Q.1   A vessel with dished ends welded on a both sides of a cylinder (90 mm dia, 15 mm thick) is to be radiographed for weld testing. Which of the technique satisfies requirements of Sample Std 1 fully?

    (a)   X rays 190 kV, DWSI, IQI 20 film Side
    (b)   X rays 220 kV, DWDI, IQI 17 film Side
    (c)   X rays 190 kV, DWDI, IQI 20 source Side
    (d)   X rays 220 kV, DWSI, IQI 20 source Side

*Ans*: For 90 mm dia both DWSI and DWDI are allowed. But, for 15 mm thickness, kV < 200 to be used. Thus only (a) and (c) remain. However Film side IQI is 17 as per the Table 14.1. Hence (a) also wrong. Thus **answer is (c)** because in any case, for DWDI the IQI has to be on the source side, which is the case here.

Q.2   What will be the minimum SFD required for the 8 mm weld to be radiographed using a source of 2.5 mmfocal spot?

    (a)   150 mm
    (b)   300 mm

**Table 14.1** IQI
Requirements

| Nominal single wall (mm) | Source side | Film side |
| --- | --- | --- |
| Upto 6.4 | 12 | 10 |
| > 6.4–9.5 | 15 | 12 |
| > 9.5–12.7 | 17 | 15 |
| > 12.7–19 | 20 | 17 |
| > 19–25.4 | 25 | 20 |
| > 25.4–38 | 30 | 25 |
| > 38–51 | 35 | 30 |
| > 51–60 | 40 | 35 |

(c)  308 mm
(d)  158 mm

*Ans*: Use **SOD ≥ 15 t²ᐟ³**. For t = 8, $t^{2/3}$ is 4 (Cuberoot of $8^2$ i.e. 64.)

SOD is 150 mm. Question is about SFD which is SOD + t = 158 i.e. **(d) is the answer**.

Q.3  What is the mandatory geometric unsharpness for a 27 mm specimen given in the Standard Sample Std 1?

(a)  f t/SOD
(b)  0.5 mm
(c)  Not specified
(d)  0.75 mm

*Ans*: The standard specifies Ug of 0.75 mm for thickness range 12–50 mm, which is applicable here. The **answer is (d)**.

Q.4  Two 15 mm thick pipes are joined using backing ring of 4 mm. The weld has 1.5 mm reinforcement. A Shim of what thickness to be kept below the IQI and which is the IQI needed?

(a)  3.2 mm shim, IQI 20 or 17 F
(b)  Shim 5.5 mm, IQI 20 or 17 F
(c)  Shim 4 mm, IQI 20 or 17 F
(d)  Shim 5.5 mm, IQI 25 or 20 F

*Ans*: As per Clause 8, whatever the shim thickness, IQI will be decided based on the parent thickness plus reinforcement, subject to maximum 3.2 mm. Here the reinforcement is only 1.5 mm. Hence IQI will be based on 15 + 1.5 = 16.5 mm. That is 20 source side or 17 film side depending upon the technique used. But to match the density of IQI with the weld, shim is needed equal to 'backing ring + reinforcement' i.e. 5.5 mm.

Thus **(b) is the answer**.

Q.5  Which of the following radiograph is acceptable for interpretation if gamma rays are used for RT?

|       | Weld density |              | IQI density |
|-------|--------------|--------------|-------------|
|       | $D_{min}$    | $D_{max}$    |             |
| (a)   | 1.9          | 3.4          | 2.8         |
| (b)   | 2.1d         | 3.7          | 3.0         |
| (c)   | 2.2          | 3.2          | 2.8         |
| (d)   | 2.2          | 2.8          | 3.6         |

*Ans*: For gamma rays min density is 2.0. Max density for x and gamma is 3.5. Thus only (c) and (d) remain. But the limit of 3.5 applies even to IQI density. Thus only **(c) is the answer**. Incidentally it also satisfies the criterion that weld to have within + 25% of IQI density.

Q.6   Which set of defects from the below is acceptable for a pipe weld 25 mm thick?

   (a)  Weld full of pores 0.5 to1 mm size, slag inclusion 7 mm long, 8 aligned pores of 1.5 mm, separated by 10 mm each.
   (b)  Wagon tracks at centre accompanied by occasional straight lines
   (c)  Slag inclusion 8 mm, distributed porosity of 2–3 mm size in 150 mm length, a pore of 4 mm size.
   (d)  15 pores of 1 mm each, a crater crack.

*Ans*: First eliminate (b) and (d) respectively for root LOF and crack. Wagon track may not be rejectable if root machining is likely to be done. But straight line with wagon tracks is LOF. (a) may be OK because for 25 mm thick weld, 1 mm pores are non-relevant indications (Table 14.2). Hence acceptable any number. Slag up to t/3 i.e. 8.3 mm is also OK. Aligned pores are OK because they are separated by 10 mm which is more than 6 L (L being 1.5 mm) So answer to be (a) if (c) is surely out. 4 mm pore in (c) is not OK as a largest single isolated pore allowed is only 3 mm. Thus **(a) is the answer**.

Q.7   A pipe with 55 mm wall thickness is tested by gamma rays. Which of the flaws below is not acceptable?

Table 14.2   Acceptance Level for random rounded indications

| Thickness t mm | Size of non-relevant indication (mm) | Maxi size of isolated indication (mm) | Distribution (Nos.)* |
|----------------|--------------------------------------|---------------------------------------|----------------------|
| Upto 10        | t/10, max 0.5                        | t/3, max 3                            | 15 of A or 2 of B or equivalent combination |
| 10 upto 30     | 1                                    | 3                                     | 4 of B or 30 of A or equivalent combination |
| 30–60          | 1.5                                  | 5                                     | 12 of B or 2 of C or equivalent combination |

(*) *Note* 1 B size = 8 of A size, 1 C size = 8 B size pores

(a)  About 30 nos of pores of 1.5 mm size
(b)  8 pores of 2.5 mm dia, one pore of 4 mm
(c)  Slag inclusions aggregating to 17 mm length
(d)  A nominal mismatch of pipes resulting in dark and light bands

*Ans*: (a) is acceptable because 1.5 mm is non-relevant indication for 55 mm job. (b) 2.5 mm is B category porosity as per Table 14.2. There are 8 of them, when 12 are allowed. But the 4 mm pore is C category and equals 8 pores of B category. Thus total 16 pores of B category against 12 allowed. Hence this is rejected. This should be the answer. (c) But yet checking further, Slag inclusion either single or cumulative length upto19 mm is acceptable. (d) Mismatch is acceptable RT defect. (Mechanical inspection should check from that angle and decide the acceptance).
**Thus answer is (b) i.e. 'Not acceptable'**

# Annexure A

Sample Std 1: 2023

## *Code of Practice for Radiographic Inspection of Pipe Welds*

(1)  Scope

This standard document deals with the requirements of the radiography inspection of welds in steel pipes up to 60 mm wall thickness.

(2)  Applicable Documents

IS 2598: Safety Code for Industrial radiography Protection.

IS 2595: Code of Practice for radiographic testing.

(3)  Safety

Adequate precaution should be taken to protect the radiography staff and other persons ion the vicinity. Provisions of IS 2598 will be followed.

(4)  Radiation Source

So far as possible X-rays will be used. Maximum kV will be as per given in the attached graph. If x-rays and gamma rays both are applicable, Gamma rays can be used if the IQI sensitivity is demonstrated (Fig. 14.1).

(5)  Film/Screens

Medium and slow speed films will be used. Lead screens will be used above 120 kV. Thickness of screen 0.05 mm till 200 kV and 0.1 mm will be used above this kV as well as for Ir-192 source.

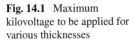

**Fig. 14.1** Maximum
kilovoltage to be applied for
various thicknesses

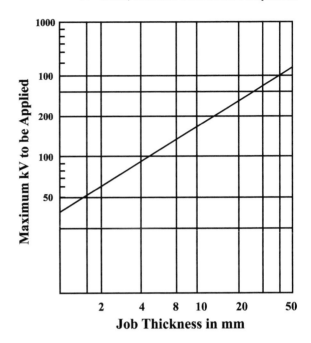

(6)   Optical Density

Minimum density on the area of interest (i.e. weld and HAZ) will be 1.8 when x-rays
are used, 2.0 when gamma rays are used. The density of the weld area can be higher
by upto 25% of the IQI density. In no case density will be higher than 3.5.

(7)   Techniques

Where ever it is possible to use Single wall technique, it will be employed. In the
'source inside' case, both, panoramic or source off-set methods are permitted.

Double Wall double image method is permitted only for dia lesser than 100 mm.
In this case either two exposures at 90° (elliptical) or three exposures at 120° are
taken.

(8)   Penetrameters

Only strip hole type penetrameter (IQI) will be used. It will be kept on the specimen
to face the source of radiation. If it is not at all possible to place the IQI on the source
side, it may be kept on the film side. In such a case it will accompany a letter 'F'
to indicate its position. Source side and film side IQI designations will be as given
below. 4T hole will be shown except for thicknesses above 25 mm, where it will be
2T.

Shims can be kept below the IQI to compensate the thickness added by the weld
reinforcement and backing strip, if any. However the selection of IQI will be based
on the parent material thickness plus permissible reinforcement i.e. 3.2 mm.

(9)  Geometry

Source to object distance should be so chosen that the geometric unsharpness remains in following limits:

Job thickness upto12 mm   0.5 mm.

Job between 12 and 50 mm thick   0.75 mm.

Job thickness over 50 mm   1.0 mm.

(10)   Acceptance

Following will be unacceptable:

 (i)  Any indication characterized as a crack or a zone of incomplete fusion or penetration.
 (ii)  Any other elongated indication which has a length greater than

    (a)   6.4 mm for 't' upto 19 mm
    (b)   1/3t for t from 19 to 60 mm inclusive

(iii)  Any group of aligned indications having an aggregate length greater than 't' in a length of 12t, unless the minimum distance between the successive indications exceeds 6L, in which case the aggregate length is unlimited, L being the length of the largest indication.
(iv)  The distribution of random rounded indications in excess of the numbers and sizes in the weld length of 15 cm will be as given in Table 14.2.

   Indications will be categorized as

A = 0.5 up to 1.5 mm.
B ≥ 1.5 up to 3 mm.
C ≥ 3–5 mm.

# Annexure B

Procedure No.RV- WP.v1.

## *Procedure for Radiographic Examination of Reaction Vessel*

1.0  Scope

This procedure describes the radiographic examination of the welds in Reaction Vessel (Component No. RV 13-IP).

Material of Construction:   SS 304/TIG welded.

No. of welds:   Eight—detailed in the Drg. No. ABCD.

Thickness Range:   4.5–56 mm.

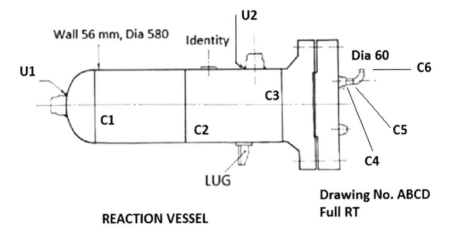

**REACTION VESSEL**

**Drawing No. ABCD**
**Full RT**

2.0   Applicable Standards

    (i)   Purchase Order XYZ-12
    (ii)   Standard Sample Std 1:2023

3.0   Qualification of Personnel

Personnel engaged in testing of this job will be those who are certified at appropriate
levels viz. Radiographers at Level-1, Supervisors at Level-2 and the Engineers at
Level-3 or as mentioned in the customer's specification.

4.0   Equipments Used

    (i)   Radiography Camera AAXX 66 with Ir 192. Source size 1.6 mm.
    (ii)   X-ray Unit MMPP 150/5 mA, focus size 1.5 × 1.5 mm

5.0   Technique Details
5.1   Exposure Geometry: As per the attached sketches (Next Page)

| Joint no | Thickness (mm) | Source | Technique | SFD (mm) | Ug (mm) | IQI |
|----------|----------------|--------|-----------|----------|---------|-----|
| C1, C2 | 56 | Ir$^{192}$ | SWSI [panoramic] | 290 | 0.38 | ASTM 35(F) |
| C3 | 56 | Ir$^{192}$ | SWSI [Source off set] | 524 | 0.19 | ASTM 40 |

(continued)

(continued)

| Joint no | Thickness (mm) | Source | Technique | SFD (mm) | Ug (mm) | IQI |
|---|---|---|---|---|---|---|
| C4, C5, C6 | 4.5 | X-ray [120 kV] | DWDI | 310 | 0.36 | ASTM 12 |
| U1, U2[a] | 56 | Ir[192] | SWSI | 300 | 0.35 | ASTM 40 |

[a]It is presumed that these welds will be radiographed early in the fabrication, when ID is accessible

*Note* Ug is calculated with largest dimension of the source as the size

Shooting geometry for representative joints in each category viz. C1, C3, C4 and U1 is given in Figs. 14.2–14.5 respectively.

5.2   Other Parameters

Film:   NDT 65 or equivalent.

Screen:   Front Pb 0.004″, back 0.010″.

IQI:   Three IQIs with suitable shim shall be kept on film side at 120° to each other during panoramic shot (C1, C2). For other cases it will be by the side of the weld.

[Note: *One can also write "as per figures"*].

Processing:   Manual processing in Kodak Developer D 198 and Kodak X-ray fixer with hardner for 5 min at 20° C.

[Note: *Alternatively, one can say " processing will be done as per procedure done as per procedure given in ASTM E 94″*].

Density:   Max. Density 3.5

Min. Density 1.8 for X-rays and 2.0 for gamma rays.
   Variation w. r. t. penetrameter − 15, + 30%

6.0   Acceptance Standards

This shall be as per ASME Sec. VIII, Division 1 clause UW 51.
   The following are not acceptable:

(i)   Any indication characterized as a crack or a zone of incomplete fusion or penetration.
(ii)   Any other elongated indication which has a length greater than

   (a)   ¼″ (6.35 mm) for 'T' upto ¾″ (19.05 m)
   (b)   1/3 t for t from ¾″ (19.05 mm) to 21/2″ (57.15 mm) inclusive
   (c)   ¾″ (19.05 mm) for 't' over 2¼″ (57.15 mm) where't' is the thickness of the thinner portion of the weld.

(iii)   Any group of aligned indications having an aggregate length greater than 't' in a length of 12t, unless the minimum distance between the successive indications exceeds 6L, in which case the aggregate length is unlimited, L being the length of the largest indication.
(iv)   Rounded indications in excess of that shown as acceptable in Appendix IV of Section VIII division I.
7.0   Report/Record

Report of observations on the films will be submitted in duplicate to the purchaser along with the films. The report will contain the technique details together with the job number and number of exposures. One copy with the purchaser's remarks will be returned to fabricator for further action.

The films will be preserved in controlled temp/humidity conditions for 5 years.

## Shooting Geometry

Panoramic: Three IQI at 120°

**Fig. 14.2**   Welds C1 and C2

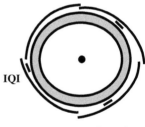

**Three IQI - Film side
120 degree apart**

**Fig. 14.3**   Welds C3

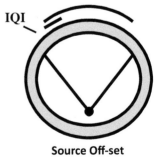

**Source Off-set**

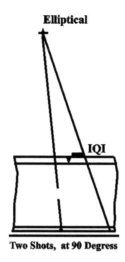

**Fig. 14.4** Welds C4, C5, C6

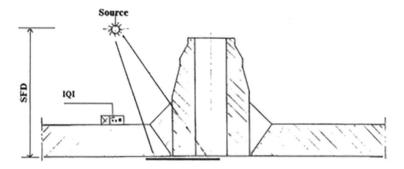

**Fig. 14.5** Nozzle welds U1, U2

| Procedure prepared by (name) | Approved by (name) |
|---|---|
| Organization | Level 3 |
| Date | Date |

# Annexure C

Procedure No. CC-Pr. v1.

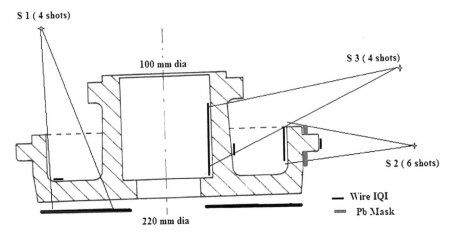

S 1 ( 4 shots)

100 mm dia

S 3 ( 4 shots)

S 2 ( 6 shots)

— Wire IQI
— Pb Mask

220 mm dia

**Fig. 14.6**   Sketch of the pump casing

## *Procedure for Radiographic Examination of Cast Casing Cover of a Pump*

### 1.0   Scope

This procedure describes the radiographic examination of the pump casing of the effluent discharge pump (Fig. 14.6).

| Material | Grade Cast Iron |
|---|---|
| As Cast thickness | As given in the sketch and in the Table 6.1 |
| Finished thickness | 3-5 mm lesser than the as-cast thickness on the faces of 2 flanges and base, where machining is involved |

### 2.0   Time of Examination

Before machining, after heat treatment.

### 3.0   Applicable Standards

    (i)   Specification XYZ-94 (purchaser's specifications)
    (ii)   ASME section VIII, Division 1 and ASME Section V, 1989 (Article 3)

### 4.0   Certification of Personnel

Personnel engaged in testing of this job will be certified at appropriate levels viz. Radiographers at level-1, Supervisors at level-2 and the Engineers at level-3.

### 5.0   Equipments Uesd

Amertest 660 Radiography Camera with $Ir^{192}$ source.
Source size 2 mm (dia.) × 1 mm (ht.).

6.0  Technique Details

Film:   Structurix D7 or equivalent.

Screen:   Front Pb 0.004″, back 0.010″.

IQI:   Source side penetrameters with designations as mentioned in the table below shall be placed:

Processing:   Manual processing in KODAK Developer D 19B and Kodak X-ray fixer with hardness for 5 min. at 20 °C.

Density:   Max. density 3.5

Min. density 1.8 for X-rays and 2.0 for gamma rays.
Variation w.r.t. penetrameter − 15, + 30%.

6.1  Shooting Geometry

| Source location | Thickness as radiographed (mm) | Source | SFD (mm) | Ug (mm) | IQI (mm) |
|---|---|---|---|---|---|
| S1 | 15 | $Ir^{192}$ | 515 | 0.06 | ASTM Wire Set B, 8th wire (0.41) |
| S2 | 24 (RT only for thick section) | $Ir^{192}$ | 520 | 0.076 | ASTM Wire Set B, 9th wire (0.51) |
| S3 | 13–17 | $Ir^{192}$ | 515 | 0.068 | ASTM Wire Set B, 8th wire (0.41) |

*Note* i. Ug is calculated with largest dimension of the source as the size
ii. If double film technique is used for two thicknesses are covered simultaneously, relevant calculation can be shown on a separate page on left side of the answer sheet to show the feasibility of such proposal

7.0  Acceptance Standards

The films will be compared to Reference radiographs of ASTM Standard E446. Category A, B and C discontinuities shall be accepted upto severity level-2. Categories D, E, F shall not be acceptable.

8.0  Repair

[Note: This clause included, if purchasers specification permits repair]

Unacceptable portions of the casting shall be repaired by welding as per the relevant clause of the spec. Repaired zone shall be radiographed with procedure commensurate with ASTM E94. Acceptance standard for the weld shall be as per UW51 of ASME Section VIII.

9.0   Report

Report of the observations on the films will be submitted in duplicate together with the job number/heat number/pattern number. Location of the source and thickness of the time of radiography shall also be mentioned. One copy of the purchaser's remarks will be returned to the fabrication for further action.

The films will be preserved for 5 years in a controlled humidity/temp conditions.

|  |  |
|---|---|
| Procedure prepared by (Name) | Approved by (Name) |
| Organization | Level 3 |
| Date | Date |

# Chapter 15
# Instruction Writing (Level 2)

## Contents

## 15.1 Introduction

The Procedure is a document connecting the Specification (or a Standard) followed during the fabrication and testing of a component with the shop floor practices. It is prepared by a person from the testing agency having level 3 certification. To give it a practical meaning, an **Instruction** is written for the shop floor operator based on this procedure. The instructions are prepared by a Level 2 person and used by a Level 1, who will carry out the testing. The Practical part of the Level 2 examination needs the candidates to write an Instruction on these lines.

## 15.2 Preparation of Instructions

Level 2 person is functionally at the Supervisory level and the operator carrying out the test would normally be a Level-1 certified person. Therefore the technique parameters (like x-ray energy etc.) will be decided by the level-2 person and the

© Ind. Society for Non-Destructive Testing 2024        141
P. R. Vaidya, *Guidebook for Radiography*,
https://doi.org/10.1007/978-981-99-8038-3_15

operator will only follow these instructions. While deciding these parameters or variables, the supervisor has one or both of the following documents to take help from:

(i)  A clause from a Code or a Standard or a specially made specification (refer Chap. 14)
(ii)  Test Procedure prepared by the Level 3 expert (refer Chap. 14)

Even when Procedure as at (ii) is available, it will be based on the applicable Specification. If the procedure given by Level 3 exists, the level 2 person does not have to necessarily look for the Specification because the essential test parameters suggested in the Procedure would be satisfying the applicable Specs.

### 15.2.1  Points to Be Covered

(The figures in bracket are the marks allotted by the ISO 9712 for the point mentioned.)

1.  Scope

Here a brief mention of the scope of the Instruction will be spelled out. It will have a number. This para will also indicate the applicable Standard or the specification. It will bear the name and signatures of the person authorizing the Instruction and the operator, with date (1).

2.  Product description

Describe the product to be tested, its material, process of manufacture (or history), identification number and dimensions. If required a sketch or drawing can be attached (2).

3.  Certification of personnel

To mention Levels required and employed (1).

4.  Equipment and accessories

Test instrument or apparatus to be specified. The details of accessories may be included. Setting up of Equipment, if any (3).

5.  Test preparations

Requirements of any surface preparation, pre- cleaning, bead grinding etc. (2).

6.  Detailed instructions for the testing (3)

Here the exact details of test procedure and the sequence of steps involved are given. In case of radiography, values of SFD, Exposure time, angle of radiation will be provided.

7.  Recording and reporting the test results $(1 + 1)$

In the case of RT, the report format will be comprehensive for full report but operator will fill some part of it like the parameters, IQI placed and zone numbers radiographed. The result part (i.e. observation and accept/reject decision) in the same form will be filled by his superiors.

This list is not exhaustive and one can include more points in the instruction depending upon the nature of job or any special demand from the specs/procedure. Important thing is to include everything necessary for the proper test. Especially for RT the following items need to be added in the Instruction. They can form part of point Nos 4, 5 or 6.

Sensitivity required; elements of IQI to be visible
Optical density on films, range and place to measure
Film selection and loading
Film identification and marking
Details of film processing
Coverage on the job and how to achieve
A statement on radiation safety.

## 15.2.2   Test Parameters

Items 4, 5, and 6 above will contain the parameters selected for testing. That also includes which option in the particular test technique to be selected. That is the important part of the Instructions and is decided by Level 3 in his Procedure; if a procedure does not exist then Level 2 needs to decide them based on the final sensitivity to be achieved. Indirectly, it will depend upon demands by the Specification. This specification or the procedure will be provided in the examination.

In the Annexures A and B, two sample Instructions are given which are based on the Procedures of Annexures B and C in Chap. 14, supposed to have been prepared by the Level 3 of the organization.

## Annexure A

Instruction No. RV-WP.v1

## *Instruction Sheet for Radiography of Reaction Vessel Welds*

1. Scope
   This is the instruction for the shop-floor operator for Weld radiography of reaction vessel—component RV.
2. Applicable Documents

(i)   The Standard SampleStd 1:2014

(ii)  Procedure No. Pr. RV-WP.v1

3. Personnel Qualification

Radiography will be carried out by AERB certified level 1 and interpretation by ISNT Level 2 personnel. Reports will be signed by a Level 3.

4. Component Description

SS reaction vessel (as per Drawing XXX) with 8 welds.

Circumferential Welds 3 Nos. (C, C2, C3), Pipe Welds 3 Nos. (C4, C5, C6) Nozzle Welds 2 Nos. (U1, U2)

Material of Construction Stainless Steel. Thickness − 4.5 to 55 mm , as per Table in Clause 6.

5. Equipment and Accessories

(i)   Radiography Camera AAXX 66 with Ir 192.

Source strength 15 Ci as on date of RT. Source size 2 mm (dia.) × 1 mm (ht.).

(ii)  X-ray Unit MMPP 150/5 mA, focus size 1.5 × 1.5 mm

Film:   NDT 65 or equivalent

Screen:   Front Pb 0.004", back 0.010"

Processing in Kodak Developer D 19 B for 5 min. at 20 °c

6. Technical Details

Shooting Geometry as per Figs. 15.1, 15.2, 15.3 and 15.4.

| Joint No. | Thickness (mm) | Source | Technique | SFD (mm) | IQI |
|---|---|---|---|---|---|
| C1, C2 | 55 | Ir$^{192}$ | SWSI [panoramic] | 292.5 | ASTM 35(F) |
| C3 | 55 | Ir$^{192}$ | SWSI [Source off set] | 530 | ASTM 40 |
| C4, C5, C6 | 4.5 | X-ray [120 KV] | DWDI | 310 | ASTM 12 |
| U1, U2* | 55 | Ir$^{192}$ | SWSI | 300 | ASTM 40 |

## *Shooting Geometry*

Three IQI at 120°

7. Records: The operator will fill out the technical details of the exposure with identification markers in the form A1. Results will be entered on the same form by Level 2 after evaluation of the films.

**Fig. 15.1**  Welds C1 and C2

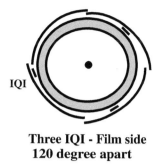

**Three IQI - Film side
120 degree apart**

8.  Report: Form A1 will be signed by Level 3. A duplicate copy of A1 will be issued to Production Section. Vessel transferred to Welding Shop.

**Fig. 15.2**  Weld C3

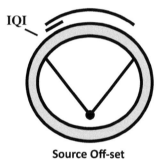

**Source Off-set**

**Fig. 15.3**  Welds C4, C5, C6

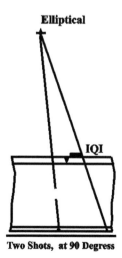

**Elliptical**

**Two Shots,  at 90 Degress**

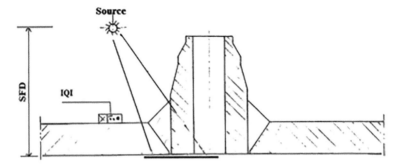

**Fig. 15.4** Nozzle welds U1, U2

| Instruction prepared by | Approved by |
|---|---|
| (Name) | (Name) |
| Level 2 | Level 3 |
| Organization | Organization |
| Date | Date |

# Annexure B

Instruction No CC- Pr. version 1

## *Instruction Sheet for Radiography of Pump Casing Cover (Casting)*

### 1. **Scope**

This is the instruction for the shopfloor operator for casting radiography of Valve Casing Cover of the effluent discharge line.

### 2. **Reference Documents**

(i)   ASME Sec VIII, Division 1; ASME Section V, Article 3
(ii)  Specification XYZ-94 (purchaser's specifications)
(iii) Procedure CC–Pr. v1

### 3. **Component Description**

Material:   Cast Iron. Component will be brush cleaned before RT.

Radiography before machining

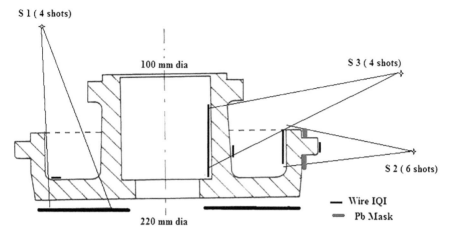

S 1 ( 4 shots)

100 mm dia

S 3 ( 4 shots)

S 2 ( 6 shots)

— Wire IQI
═ Pb Mask

220 mm dia

**Fig. 15.5** .

As Cast Thickness:   As given in the sketch (See Fig. 15.5).

### 4.  Personnel Qualification

Radiography will be carried out by the BARC Certified Level 1 radiographer. Under instructions from level 2.

### 5.  Equipment to be Used

Radiography Camera AAXX 66 with $Ir^{192}$ source.
Source strength 15 Ci as on date of RT. Source size 2 mm (dia.) $\times$ 1 mm (ht.).

### 6.  Technique Details

| | |
|---|---|
| Film: | Structurix D7 or equivalent |
| Screen: | Front Pb 0.004", back 0.010" |
| IQI: | Source side penetrameters with designations as mentioned in the Table |
| Identification Markers: | For shots S1, four zones A-B-C-D-A will be marked along with letter S1 every time. Similarly 6 zones for S2 and 4 zones for S3. |
| Processing: | Manual processing in KODAK Developer D 19B and Kodak X-ray fixer with hardness for 5 min. at 20 °C. |
| Density: | Max. density 3.5, Min. density 1.8 for X-rays and 2.0 for gamma rays |

## 6.1  **Shooting Geometry**:

| Source location | Thickness as radiographed (mm) | SFD (mm) | Exposure (Ci hrs) | IQI |
|---|---|---|---|---|
| S1 | 15 | 515 | 1.8 | ASTM Wire Set B, 8th wire (0.41 mm) |
| S2 | 24 (RT only for thick section) | 520 | 2.8 | ASTM Wire Set B, 9th wire (0.51 mm) |
| S3 | 13–17 | 515 | 1.6 | ASTM Wire Set B, 8th wire (0.41 mm) |

7.  Record/Results

The operator will fill out the technical details of the exposure with identification markers the form A1. Results will be entered on the same form by Level 2 after evaluation of the films.

8.  **Report**

Form A1 will be signed by Level 3. A duplicate copy of A1 will be issued to Production Section. Specimen handed over to Machining shop.

| Instruction prepared by | Approved by |
|---|---|
| (Name) | (Name) |
| Level 2 | Level 3 |
| Organization | Organization |
| Date | Date |

# Useful Formulae in Radiography

## Basic Physics of Radiation

Change in radioactivity with time $\quad N = N_0 \exp^{-\lambda t}$

Change in Radiation Intensity $\quad I = I_0 \exp^{-\lambda t}$

Half Life $\quad T_{1/2} = 0.693/\lambda$

Inverse Square Law $\quad I_1/I_2 = D_2^2/D_1^2 \ \text{ or } \ I_1 D_1^2 = I_2 D_2^2$

## Radiographic Equipments

Intensity of a X-ray source $\quad I = C\,i\,Z\,V^2$

Specific Activity of isotope source $\quad S = \lambda\,N_0/A \cdot Bq/g$

## Interaction of Radiation with Matter

Exponential Absorption $\quad I = B\,I_0 \exp(-\mu x)$

Build Up Factor $\quad B = 1 + I_s/I_d$

Half Value Layer $\quad HVL = 0.693/\mu$

Mass Attenuation Coefficient $\quad \mu_m = \mu_L/\rho$

## Radiographic Films and Screens

Optical density $\quad D = \log_{10} I_d/I_t \quad \text{or}$

$$D = \log_{10}\left[\frac{1}{transmittance}\right]$$

Film Gradient $\quad G = d(D)/d(\text{Log rel Expo})$

## Basics for Radiography Techniques

Direct Square Law for Exposure $\quad \dfrac{E_2}{E_1} = \dfrac{D_2^2}{D_1^2}, \quad$ for Exp time $\dfrac{t_2}{t_1} = \dfrac{D_2^2}{D_1^2}$

Exposure time (h) $\quad \dfrac{Film\ Factor \times SFD^2 \times 2^{\frac{t}{HVL}}}{Ci \times RHM} \quad$ SFD in metres

Geometric Unsharpness $\quad U_g = \dfrac{f \times OFD}{D_0} \text{ or } \dfrac{f \times t}{D_0}$

Direct Enlargement $\quad U_g = f(M - 1)$

© Ind. Society for Non-Destructive Testing 2024
P. R. Vaidya, *Guidebook for Radiography*,
https://doi.org/10.1007/978-981-99-8038-3

## Quality of Radiographic Image

Equivalent Pene. Sensitivity   $EPS (\%) = \frac{100}{x}\sqrt{\frac{Th}{2}}$

Per Cent Contrast                  $\frac{\Delta x}{x} \times 100 = -2.3\Delta D/x\mu G_D$

## Special Applications and Techniques

Depth of Flaw (Parallax method)   $d = D\Delta F/(\Delta S + \Delta F)$

Location   of   Flaw   (Double   $h = t \times \frac{(\Delta f - \Delta BM)}{\Delta TM - \Delta BM}$
Marker)

# Answers to the Model Questions

**Basic Physics of Radiation**

Q.1    (c)      Q.2  (c)      Q.3  (b)

**Radiographic Equipments**

Q.1    (d)      Q.2  (c)      Q.3  (b)      Q.4  (d)      Q.5  (d)

**Interaction of Radiation with Matter**

Q.1    (b)      Q.2  (b)      Q.3  (b)

**Radiographic Films and Screens**

Q.1    (c)      Q.2  (b)      Q.3  (b)

**Basics for Radiography Techniques**

Q.1    (d)      Q.2  (b)      Q.3  (c)

**Quality of Radiographic Image**

Q.1    (c)      Q.2  (d)      Q.3  (d)

**Radiography Application and Techniques**

Q.1    (b)      Q.2  (b)      Q.3  (d)

**Interpretation of Films**

Q.1    (a)      Q.2  (c)      Q.3  (d)      Q.4  (a)

**Advance Techniques**

Q.1    (c)      Q.2  (b)      Q.3  (c)

**Special Applications and Techniques**

Q.1    (b)      Q.2  (a)      Q.3  (c)

**Filmless Options and Image Processing**

Q.1    (d)    Q.2  (c)    Q.3  (b)

**RT in Different Industrial Sectors**

Q.1    (b)    Q.2  (c)

**Radiation Safety**

Q.1    (b)    Q.2  (b)    Q.3  (d)

# Index

© Ind. Society for Non-Destructive Testing 2024
P. R. Vaidya, *Guidebook for Radiography*,
https://doi.org/10.1007/978-981-99-8038-3